A Bio-Bibliography for Biruni

Abu Raihan Mohammad Ibn Ahmad (973–1053 C.E.)

M. Kamiar

The Scarecrow Press, Inc.
Lanham, Maryland • Toronto • Oxford
2006

SCARECROW PRESS, INC.

Published in the United States of America
by Scarecrow Press, Inc.
A wholly owned subsidiary of
The Rowman & Littlefield Publishing Group, Inc.
4501 Forbes Boulevard, Suite 200, Lanham, Maryland 20706
www.scarecrowpress.com

PO Box 317
Oxford
OX2 9RU, UK

British Library Cataloguing in Publication Information Available

Library of Congress Cataloging-in-Publication Data

Kamiar, M., 1946–
A bio-bibliography for Biruni : Abu Raihan Mohammad Ibn Ahmad (973–1053 C.E.) / M. Kamiar.
p. cm.
Includes bibliographical references.
ISBN-13: 978-0-8108-5663-9 (pbk. : alk. paper)
ISBN-10: 0-8108-5663-8 (pbk. : alk. paper)
1. Biruni, Muhammad ibn Ahmad, 973?–1048—Bibliography. I. Title.

Z8098.7.K375 2006
[Q143.B5]
509.2—dc22 2006001890

The paper used in this publication meets the minimum requirements of American National Standard for Information Sciences—Permanence of Paper for Printed Library Materials, ANSI/NISO Z39.48-1992.
Manufactured in the United States of America.

Table of Contents

Preface

I heard of Biruni's name when I was only eleven years old in the fifth grade in the elementary school in my hometown of Bijar in the Iranian province of Kurdistan. Biruni's name came up in a history class in passing. We learned that he was only one of the Muslim geographers. We even had to memorize the names of some of his books. Nevertheless, we saw him as one of many stars in the Iranian-Islamic scholarship. For many years, during my high school, and teachers training college, Biruni was a fading star in a distant sky.

In 1963, I became a teacher when I was only seventeen years old. In 1970, I passed the "College Entrance Examination" and became a geography student in the University of Esfahan. With a Bachelor Degree in geography, I graduated from this university in 1974. I still did not know much about Biruni. He simply was another name on a long list of famous Persian scholars.

It was after my arrival in the United States to continue my studies in geography that I had another encounter with Biruni. As graduate students at Michigan State University, we were required to take a course called "GEO 852: Philosophy of Geography". I took this course with Professor Lawrence M. Sommers in the fall of 1979. As a requirement in this course, we had to write a short, five-page, paper about a geographer of our choice. I selected Abu Raihan Biruni. My short report was a formal research paper. It contained another three pages of footnotes and references. Professor Sommers liked my paper very much. In his comments, he used the word "excellent" and suggested that that my paper was publishable. It was after this encouragement that I started to collect all of the references on Biruni.

References on and about Biruni are abundant but badly scattered. These sources are found in many different languages, sizes and cover varieties of subjects that beg to be combined and put together under one title. None of the bibliographies available to us had a complete list of Biruni books. More importantly, they need to be updated and integrate the newest technology in library research, the Internet.

Thus, the goal of this selected research bibliography was to introduce the writings and collect all references associated with one of the greatest scholars that our world has seen, namely: Abu Raihan Mohammad Ibn Ahmad Biruni (973-1053 C. E.). It is hoped that this project present to the western world a bright star in the field of scientific achievements and rekindle interest in those who already knew about Biruni.

I would like to acknowledge the generosity and guidance I received from both Professor Seyyed Hossein Nasr of George Washington University and Professor EhsanYarshater of Columbia University. I also would like to express my appreciation to Marwan Abu-Sawwan, Professor of Mathematics at my college, who helped with correcting and translating the Arabic version of Biruni's books. Dr. Nancy Yurko, the Dean of our program, provided us with an office, a computer, and some financial help to cover the cost of the Inter Library Loans.

The staff of the Library of Congress was most helpful to me during my week spent carrying out research in the African and Middle Eastern Reading Room. I also appreciate the help that I obtained from the library at Florida Community College at Jacksonville. Indeed, it would not have been easy to finish this book without help from the librarians here. Ms. Barbara Markham's expertise and assistance in securing many inter-library loans is much appreciated, as is Ms. Barbara Salvage's research and editorial assistance. However, I assume responsibility for any error in this bibliography. Any problem in this collection should be referred to me.

Dr. Mohammad S. Kamiar

Jacksonville, Florida

Introduction

This project was undertaken because no complete or up to date bibliography on one of the most famous scholars in the history of mankind was available. The three existing bibliographies on Biruni were all either old or incomplete and begged for updating. Our bibliography was prepared as a response to this need. This selected research bibliography was prepared for those individuals who are interested in the history of science, scientific research, and scholarship.

The goal of this research was to collect and assemble in a single volume all published works in English and Persian, related to Abu Raihan Biruni, an Iranian scholar who lived between 973 and 1053 C.E. Here, for the first time, the complete names of all 183 books by Biruni are provided as well as documentation of Internet sites dealing with him and his works. In addition to the traditional library research, we have inspected the Internet and the most important Web sites were identified and listed. We were also able to document the existence of a full-length film based on Biruni's life, as well as a doctoral dissertation. Thus, our main goal was to facilitate further research on Biruni's life and works.

Several bibliographies on Biruni have already been published in different languages. We had access to only three of them. The first two are in English and the third is in Persian. The first bibliography that we found was completed by Nasr in 1973. Another bibliography was collected by Azkayi was also published in 1973. This *Catalogue* is in Persian and has a list of only 180 of Biruni's books. Khan compiled the third bibliography of Biruni's work in 1982. Of course, all three of them are old and beg to be updated. Actually, Nasr (1973) hoped that "others will complete and bring it [his bibliography] up to date in the future" (p. 9). And, we have accepted this challenge. However, because of technological achievements, we went a step further by investigating the Internet. Yet, we do not claim that our bibliography is definitive, exhaustive or complete. Indeed, we avoided any literature and references in languages other than English and Persian. Even for these two languages, we may have unintentionally omitted a few sources of which we were not aware.

Our selected research bibliography has five main parts. In the first part called "General Background", we introduced Biruni by dealing with his time, his world and his life. The second part includes full names of 183 books written by Biruni. The titles of these books are given in Arabic, Persian, transliteration of the Arabic title, and English. All of these books are annotated and if available the number of folios is given for each one. This part contains the most complete list of Biruni's books. Utilizing this format our selected research bibliography gathers all available information under one heading for the first time. In part three, we listed available references related to Biruni. This part has four sections: articles, books, dissertation, and film. In part five, we attempted to provide the reader with a study of selected Internet-accessible references on Biruni. In the beginning of each section we have furnished a brief explanation of our approach and methodology.

Biruni was more than just a great scholar. He was a humanitarian individual and a man of honor. He was the first scholar to predict the existence of landmasses on the other side of the planet Earth, now called the Americas. He preceded the European Renaissance by five centuries. If he were alive today, it is likely that his research agenda and achievements would have won Nobel prizes in many fields.

We hope that we have succeeded in communicating to our readers the exceptional achievements of this extraordinary scholar.

Part I

General Background

Abu Raihan Biruni

His full name was Abu Raihan, splled also as Abu Rayhan, Mohammad Ibn Ahmad Biruni. He was born in 973 C. E. in a suburb of the city of Kath, now called Khiva, today in the country of Uzbekistan in central Asia. He died sometimes after 1050, most likely in 1053, in Ghazna, now called Ghazni, in Afghanistan. Biruni is said to be one of the greatest scholars that ever lived on Earth. To many people, especially his students, he bore the title "Ostad", which translates into English as "the Master." In this section Master Biruni will be introduced. This introduction deals with his time, his world, and his life achievements.

His Time:

Biruni's birth date millenary was celebrated in 1973 in several countries including Afghanistan, India, Iran, Uzbekistan, and Pakistan. Thus he lived in a time that is known as the Middle Ages. This period of European history is called the Dark Ages. The Moslem world, however, was not so "dark" during the Medieval Ages. In fact, many scholars were at work doing research on every aspect of life. The Moslem world was large and rich. It included Southern Europe, North Africa, the Middle East, and Central Asia. Cordova, Cairo, Baghdad, Rayy, Gorgan, Isfahan, Khwarazm, Bukhara, and Ghazna were among the most important centers of science, civilization, and culture of that era.

The golden age of Baghdad began in the early ninth century. It had the largest number of universities, libraries, observatories, and hospitals in the world. At one time, Baghdad's Academy of Wisdom had seventy scholars as full time members. They were paid handsomely to work on the translation and dissemination of knowledge from other parts of the world including Greek-Roman, Persian, and Hindu classical research.

Moslems, however, were not just keepers of the world's knowledge. They themselves carried out independent and critical research and improved on what was available. Biruni, therefore, was fortunate to be able to utilize a fruitful and maturing age of scientific achievement. Many important scholars were his immediate predecessors and contemporaries. Biruni was familiar with and studied books written by Abu Nasr Farabi, Mosa Khwarazmi, Aul Wafa Buzjani, Ibn Yunus, Al-Bakri, and Ibn Hazm. He had correspondence and debated with Avicenna, Ibn Sina, a Moslem physician-philosopher who is well known in the West. Biruni met Ferdosi, who is considered to be one of the best epic writers in the world.

By the end of the first millenium, research and scholarship slowed down in the Moslem world. The caliphs degenerated into a symbol of corruption, luxury, and immorality selling honorific titles to whoever could afford to pay. Many Persian scholars such as Mosa Khwarazmi, Muhammad Zakariya Razi, Abu Wafa Buzjani, Abdul Samad, Abu Sahl Masihi, Abu Hassan Khammar, and Abu Nasr Mansur tried to reinvigorate research that had fallen into a state of stagnation. Most of these scholars came from the extreme eastern part of the Moslem world and most of them lived in the Persian kingdom of Khwarazm.

The biggest shift in research came after a major debate between Ibn Sina and Biruni. By questioning Ibn Sina's research method, Biruni aimed to free scientific research from the classical Greek system particularly that of the Aristotelian approach. This debate was a turning point in the history of scientific studies. Biruni promoted a movement away from simple translation of Greek texts, towards a different research procedure, involving investigation and experimentation, in a format that celebrates independent critical thinking and that was inductive rather than deductive in nature.

As mentioned before, Moslem scientists functioned not only as the custodians of what they borrowed and translated into Arabic, but also they learned Greek scientific theories and applied them as working hypotheses requiring further research in the field. Moslem scientists in search of truth tested those hypotheses, accepted some, added new data, and improved methods of scientific investigation. The European Renaissance was also fueled by translations of over 500 years' worth of Moslem scientists' works in science and technology into European languages. We know some of Ibn Sina and Razi's books were translated into Latin in the late 13th century and used as textbooks in many European universities and institutions of higher education. It is believed that without help from the Middle East, it would have been difficult for Europe to achieve what is called the Renaissance and the Age of Enlightenment.

His World:

Abu Raihan Biruni lived most of his life in the two cities of Kath on the eastside of Amu Darya River and Gurganj on the opposite bank. Both of these cities were located south of Lake Aral in the kingdom of Khwarazm, now in Uzbekistan, Central Asia. At one time or another, one or both of these two cities were called Khwarazm and were the seat of the government, sometimes simultaneously. The Greeks called this region Transoxania. The Arabs named all of the lands beyond Amu Darya River, Mawara al-Nahr, which included Khwarazm, and, now is generally referred to as Central Asia, mostly located in the modern country of Uzbekistan.

Khwarazm was situated in the extreme northeastern corner of the Moslem Empire. With a distance of over a thousand miles from Baghdad, the region was also the farthest from the center of control. On two sides, north and east, this region bordered on non-Moslems. Yet, Khwarazm benefited from its mere location. It was isolated from the center of control, orthodoxy, and hypocrisy. At the same time, it was exposed to Christianity, Hinduism, Buddhism, and Zoroastrianism that was still practiced in the region. This resulted in relative freedom of thought and independent research by scholars in the periphery of the Moslem world.

As a frontier land, Khwarazm was not far from Turkistan and Mongolia. This region was and still is surrounded by deserts on all sides, except from the south. Since the early times, waves of Turkish and Mongolian nomads entered the region in search of grasslands. Although the most destructive invasion appeared in the 13th century, the region has always been at the mercy of nomadic tribes. This is the same century when Europe was entering its Renaissance.

Khwarazm was a rich kingdom. Located at a safe distance from the silk route; urban areas, especially Kath and Gurganj, prospered due to international commerce and suitable farming areas within the kingdom. Prosperity of the rural areas was based on availability of fresh water from Amu Darya River, fertile soil, ample sunshine, and human ingenuity. Here hard working rich farmers produced every product and raised animals on their farms. Fish and sea products as well as mineral resources such as turquoise, ruby, and gold were exported from Khwarazm. Manufactured goods such as boats, locks, silk and cotton cloths, fabrics, carpet, blankets, satin for royal gifts, wool products, perfume, musk, and soap, and many other luxury products were typical export items from Khwarazm.

The most extensive irrigation canals in Central Asia were found in Khwarazm. These canals were used for irrigation, transportation, and redistribution of excess water due to the flooding of Amu Darya. According to Barthold (1968, p. 144) some of these life-giving canals had impressive dimensions. For example, the Madra canal was about 24 feet wide and 20 feet deep. These canals were indications of human ingenuity, prosperity and determination to live in harmony with nature. The Russians have discovered and unEarthed some of these irrigation canals.

Life in Central Asia would have been almost impossible without help from the rivers of Syr Darya and Amu Darya that originate in the mountainous regions in Afghanistan. As an oasis in the middle of deserts, Khwarazm was a product of Amu Darya River. This river can be referred to as the Nile of Central Asia, because of its life-giving importance to this region. We have historical reports that some fresh fruits like melons, grapes, berries, and pomegranates from Khwarazm found their way into bazaars of Baghdad. Melons and other fresh fruits were exported in leaden snow-packed containers to destinations as far as the heart of the Islamic Empire over a thousand miles away. The use of snow in transportation of fresh products over such a distance during summer months elucidates the smart economic activities of this part of the world in medieval times. Rich merchants from Khwarazm, with their expensive pointed fur hats, were seen in many big cities all over the world.

Farmers in Khwarazm were mostly wealthy landowners. Cities were centers of manufacturing activities and trade. In fact, trade made this kingdom very prosperous. Thus, government was rich and could afford to provide public services to the people. The general population in Khwarazm was much better off and more educated than the rest of the world. This is where our hero grew up. This mostly autonomous, rich isolated frontier oasis could afford to have many educational facilities. In return, it has produced many scholars with the highest professional standards, among them Abu Nasr Mansur and Abu Raihan Biruni.

However, the kingdom of Khwarazm, suffered not only in the hands of invaders but was also devastated by its own unique geography. These contributing factors included accumulation of salt on the farms in the delta area, periodic floods, Earthquakes and epidemic diseases such as cholera and typhus. The two major cities of Khwarazm were destroyed several times by flood. Both of them have been relocated several times. At some future date, after Ptolemy prepared his world map, Earthquakes hit most of the areas east of the Caspian Sea. We do not know how many people were killed, but we know these quakes changed the bed of Amu Darya. Ptolemy's map showed this river emptying its water in the Caspian Sea. From Biruni's life to the present day, Amu Darya flows north towards Lake Aral. As the result of this change the east bank of the river flooded and the areas towards the Caspian Sea turned into some steppes and mostly desert.

Large cities in Central Asia became important centers of commerce with China. Khwarazm was a stepping stone for the diffusion of Chinese inventions westward. The three Chinese inventions, compass, paper, and gunpowder, that changed the history of mankind had to pass through Central Asia. The first two were used extensively in the Moslem world. The city of Samarqand and later Baghdad became important paper producing centers. Gunpowder was re-invented by Abu Nasr in Khwarazm. As a prince-scholar, Abu Nasr was Biruni's teacher, mentor, and colleague. Gunpowder was then stolen and brought to Spain and later used to defeat the Moslem Moors in 1492. In fact, European recovery from Moslem rule corresponds with the discovery of the New World and their Renaissance in 1492, the most important year in the history and geography of mankind.

His Life:

Around the turn of the first millennium lived a scholar, a genius of the highest magnitude. As a child he was named simply Mohammad. In elementary school he was usually referred to as Ibn Ahmad, son of Ahmad. His full name evolved slowly. As a reference to his place of birth, he was called Biruni. Meaning "from outside" in Persian, Biruni was actually a derogatory name imposed on him due to the fact that he was not born inside the city proper. In the history of mankind he is the only person so nicknamed. He took this intended insult and used it to indicate that although he was not a member of the rich aristocracy, he had critical outsider perspective and an insightful view of the world around him. When just a young scholar, he had to select a nickname for himself. He chose Abu Raihan. "Raihan" means "Sweet Basil." Lady Raihanah was the name of the girl he loved. Therefore, his full and formal name became *Abu Raihan Mohammad Ibn Ahmad Biruni.*

Biruni was born on the 4th of September 973 C. E. After having an eventful, fruitful and memorable life, he died in Ghazna, now called Ghazni, located in Afghanistan, sometime after the year 1050. He lived to be about 80 years of age, quite an achievement in itself.

He was almost fourteen years of age when he was taken into slavery and sold to the "White Dressed" people of Tammami to be sacrificed. A brave prince-scholar by the name of Abu Nasr saved his life. Following this unnerving experience, life for Biruni focused on studying, research, and writing. Most of his life's work dealt with different aspects of the sciences, particularly geography.

After Abu Nasr rescued him. Biruni's life, like other people, was a struggle for survival. However, this man of honor struggled in the best possible way. Although he had to leave Khwarazm and lived in poverty in the city of Rayy, now a suburb of Tehran, he became the most respected scholar in Gorgan, southeast of the Caspian Sea and Ghazna, Afghanistan. He became very sick in India and many times he complained about lack of income, facilities, and opportunities to carry out research. Fortunately, he survived, aged gracefully and is known as a genius of the first magnitude.

He worked every day of the year, save only two equinoxes, the first day of spring and autumn. He wrote a monumental number of books and treatises. We have names of 183 books and treatises associated with Biruni. According to many historical documents, he wrote over 13,000 folios; a folio is about a page of written words. This is estimated to be the full load a strong camel can carry. One of the largest number of pages output by a single person in history, this is a formidable achievement for his time because those books and treatises consisted largely of highly technical materials in almost every scientific field from many different sources. Unfortunately, only 13 of those books and treatises have survived. Regretfully few people around the world, especially in the West, know this "one of a kind scholar", a man with a very powerful and beautiful mind.

Since youth, Abu Raihan had a passion for learning objective knowledge. The pursuit of this noble goal motivated his desire to learn foreign languages. He obliged himself to go directly to the original sources and ethnic groups. He learned at least eight different languages and utilized them in his research. Although his mother tongue was Khwarazmi, his major books and treatises are in Arabic, the lingua franca of his time. He also knew Greek, Hebrew, and Syriac and Turkish. He wrote a few of his books in Persian and his research in India could not have been accomplished without his knowledge of Sanskrit.

Master Biruni was what we call a "Jack of all Trades". A first-rate researcher, unmatched scientist, and original thinker, he covered almost every discipline in the annals of human intellect. He was the foremost astronomer-mathematician-historian-geographer of all time. His areas of interest encompassed all disciplines from anthropology to zoology, alphabetically speaking. He wrote about anthropology, arts, astronomy, botany, cartography, chemistry, chronology, civilization, ethnography, geography, geology, geomorphology, geodesy, medicine, mathematics, metaphysics, mineralogy, optics, physics, philosophy, physiology, religion, sociology, survey, and zoology. In addition, he authored comedy, fiction, and poetry.

With a list of 183 books and over 50 different inventions, Master Biruni achieved his goal of objective reasoning. His inductive research method resulted in many practical scientific applications. His critical method very much resembled that of modern scholars. Biruni's outstanding qualities in research include his investigative curiosity in the area of the applied sciences that he carried out by consulting multiple original sources without prejudice. His objectivity and continual search for scientific truth make him a model scholar for all times.

Biruni was nearly twenty years of age when he wrote his first book. His research continued until he reached the age of eighty. During these years he produced 183 books of the highest quality. His book, *the Chronology of Ancient Nations*, is the best, probably the first, and one of the few books written on the subject of "time". Another one of his books, *Masud's Canon*, is one of the most important books in astronomy. His *Book of India*, among the last books he wrote, is still seen as one of the best sources of information about this country and it's civilization. And, moments before he died, he was thinking about a mathematical formula on the laws of inheritance.

Biruni believed that our planet was spherical in shape. He was aware of the Earth's movements. He is the first scholar who calculated a nearly correct size of the circumference of our planet. Biruni believed that our planet was millions and millions of years old. He knew about landmasses opposite to Eurasia. He is the first and the only scholar that made a reference to the Americas 500 years before Columbus.

He had several important scientific ideas that are similar to modern concepts. Like Malthus, Biruni blamed destruction of the ecosystem on uncontrolled population growth. Biruni argued the nature's work was like a gardener and the strongest and fittest could survive. He thus argued for some sort of basic Darwinian theory of natural selection and evolution. His geological ideas did not differ from modern concepts either. Biruni hypothesized that an ocean covered India and Arabia during bygone years. Geologists nowadays call this Mesozoic Ocean the Tethys.

Biruni's ideas reached Europe in a less formal way. It is highly likely that men like Prince Henry the Navigator, Christopher Columbus, Amerigo Vespucci, Vasco da Gama, Magellan, Copernicus, Galileo, Newton, Bacon, Malthus, and Darwin somehow benefited from Biruni's writings and research findings. Is it coincidental that Galileo's telescope could be seen as merely an improved version of Biruni's astrolabe? Or, is it by chance that gravity equations formulated by Biruni and Newton are virtually the same?

Edward C. Sachau (1845-1930), a German professor of Oriental Languages in the Royal University of Berlin was more familiar with Biruni's works than any other scholar. Sachau translated two of Biruni's books, the *Chronology of Ancient Nations* and *Book of India* into German and English. Sachau believed that Biruni's work is "like a magic island of quiet, impartial research in the midst of a world of clashing swords, burning towns, and plundered temples" (1887, p. xxiii). It is no wonder that George Sarton called the first half of the 11th century the "Age of Biruni" (1927, p. 693). He believed that Biruni was an adventurer and a discoverer. To him, Biruni "may be counted as one of the greatest geographers of all times, but he was alone" (p. 699).

Regretfully, Biruni's achievements of the highest rank have escaped general recognition. One can not afford to allow citizens of the world to ignore the past, impoverish the present, and endanger the future. We shall remember Biruni as one of the torchbearers of our intellectual history.

Part II

Names of 183 Books Written by Biruni

Three sections are included in this part. First, some methodological remarks are presented. Here we provide the readers with information on our references, languages we selected, and classification of Biruni's 183 books. In the second section we introduce two different classifications of Biruni's works. Our general observations on some of the most important repeated keywords in the titles of this great scholar's books are discussed in this section. The third section represents, for the first time, the most complete list of all 183 books written by Biruni. Each entry furnishes the reader with the name of a book in Arabic, in Persian, an approximate transliteration, in English, with a brief annotation.

Methodology:

We had access to the titles of 183 books bearing Biruni's name. He himself, in a book named *Risala fi Fihrist Kutub Muhammad Ibn Zakariyia al-Razi*, translated as "*On the List of the Books of Muhammad Ibn Zakariyai al-Razi*" has documented most of these titles, 138 of them. The short form of this work's name is the *Fihrist*. Please see entry number 168 on our list. Collected by Biruni, in the *Fihrist* one finds the names of all of the books written by Muhammad Ibn Zakariyai Razi (865-925 C.E.), who was a celebrated alchemist-physician-surgeon-philosopher. He was born and died in the city of Rayy, not far from the village of Tehran. In addition to listing all of Razi's books, Biruni provided a list of his own writings. Thus, the *Fihrist* is considered to be one of the first bibliographies and the most reliable source of information about the works of Razi and

Biruni. It serves as the first reliable reference for Biruni's writings. Thus all succeeding bibliographies of Biruni's works were based on the *Fihrist.*

Likewise, different historians mention some of the following books. In a few cases several chapters from Biruni's books are copied in books written by historians and scholars after his death. Even, at least in one case, a whole book was reproduced. For example, Biruni's *History of Khwarazm* is entirely reproduced by Bayhaqi (please see number 160 on our list). Unfortunately, the great majority of these books are lost.

Some of his colleagues wrote books and asked Biruni to list them under his name. Thus, it is important to note that not all of these 183 books were written by Biruni himself. In the *Fihrist,* Biruni provides the readers with a list of books composed by Abu Nasr, Abu Sahl Masihi, and Abu Ali Hassan Ibn Ali Jili. Please see the last three lines in the following table of Biruni's classification. Abu Nasr Mansur Ibn Ali Ibn Iraqq was a member of the royal family of the kingdom of Khwarazm. He was Biruni's teacher, colleague, and friend. Biruni calls him his "Master." We do not have much information on Abu Ali Hassan Ibn Ali Jili. But, we know that Abu Sahl Masihi was a colleague of Biruni. As a bilingual Christian scholar, Abu Sahl helped Biruni in his Greek studies. Together they also studied the reasons for extreme temperatures and several other scientific subjects.

Khan's (1982) Bibliography provides full information on manuscripts, editions, translations, and articles written on most of Biruni's books. Therefore, we found it unnecessary and redundant to repeat or update. We, however, provided full information on those books that have been translated into English. Full information about available manuscripts has been included. Interested readers may consult pages 11-57 in Khan's Bibliography.

In the present bibliography, the titles of Biruni's books are given in the following order:

1) Arabic script

2) Persian translation

3) Approximate transliteration, and

4) Meaning of the title in English.

Although it was not the earliest bibliography of Biruni's works, Ahmad Saeed Khan's Bibliography is the most comprehensive and the most recent. It provided us with the full names of Biruni's 183 books. We used Khan for identifying English and Persian articles and books about Biruni. We added the Arabic scripts, Persian translations, and our annotations to this list. Thus, the above arrangement, providing for multiple possible ways of presenting a single book, was carried out for the first time in this bibliography.

In addition to Khan's Bibliography we had access to two other major sources, both published on the occasion of Biruni's millenary celebration in Tehran in 1973. The first is *Karnamehaya Biruni: Bibliography and Catalogue of His Books*, compiled by Professor Parviz Azkayi. This book is in Persian and has a list of only 180 of Biruni's books. The titles are given in their original Arabic forms with the Persian translations. Our Arabic titles were virtually identical with those of Azkayi. We compared our Persian titles with those in Azkayi's *Catalogue*, and the agreement was remarkable. The second major source is Dr. Seyyed Hossein Nasr's *Al-Biruni: an Annotated Bibliography*. Nasr does not list all of Biruni's books or their contents but instead focuses on listing references about Biruni's works by other scholars.

Khan and Azkayi's bibliographies are very similar. Although Khan's Bibliography was published ten years after Azkayi's Catalogue, we were unable to find reference to the latter in Khan's work. It is surprising that Khan was unaware of Azkayi's book. These two references vary only slightly as to the number of folios in Biruni's books. Yet, these numbers are very similar. The most notable difference we observed concerned Biruni's first book: Azkayi's number of folios for this book was 250 and Khan's number was 1250. Thus, to indicate this problem we put a question mark in front of the latter number.

Classifications and General Observations about Biruni's Books:

Biruni's books are not listed alphabetically. Instead, the list is based on the main subject of each book. Actually, Biruni himself in the *Fihrist* has classified the first 138 books into 16 categories that are shown in the following table.

Table 1
Subject List of Biruni's Works
(Works Classified by Biruni Himself in *Fihrist*)

I.	Astronomy (Notions, Tables, Calculations)	1-18
II.	Longitudes, Latitudes, Azimuths (Geography)	19-33
III.	Arithmetic (Numeration, Calculation)	34-41
IV.	Rays and Passing (Astrological Notions)	42-45
V.	Instruments (Astrolabe)	46-50
VI.	Time measurement (Chronometer, Chronology)	51-55
VII.	Comets (Flowing Stars)	56-60
VIII.	Miscellaneous (Astronomy, Geometry, Geography, Physics)	61-72
IX.	Astrology (Principles, Horoscopes)	73-79
X.	Anecdotes (Translations of stories from Persian or Sanskrit)	80-92
XI.	Customs and Religion	93-98
XII.	Miscellaneous (MSS lost by the author)	99-103
XIII.	Miscellaneous (After attaining the age of 61 years)	104-113
XIV.	Miscellaneous (Works composed by Abu Nasr)	114-125
XV.	Miscellaneous (Works composed by Abu Sahl)	126-137
XVI.	Grammar (Work composed by Abu Ali al-Jili)	138

Khan (1982), on the other hand, has divided Biruni's 183 books into seven main categories with some sub-categories. Khan's classification is especially useful for identification of those books, 45 of them (number 139 to 183), that were not mentioned by Biruni himself. Table 2 shows Khan's classification of Biruni's 183 books.

Table 2
Khan's Classification of Biruni's Works

I.	Mathematics:	
	(1) Arithmetic	34-41
	(2) Geometry	64, 66-67, 108, 125-126, 135, 167, 175, 178
	(3) Trigonometry	124, 180
	(4) Stereography	48, 110, 139, 142
	(5) Survey or Measurement of Land	65
II.	Astronomy and Astrology:	
	(1) General Astronomy	1-18, 61-62, 71-72, 100-101, 104, 106, 111, 113, 115-118, 120-121, 123, 127-128, 137, 140-141, 143-147, 155, 171, 176, 179, 181
	(2) Meteors	55, 57-60, 69
	(3) Instruments	46-47, 49-52, 107, 119, 122, 169, 177
	(4) Astrology	42-45, 70, 73-79, 99, 102, 132, 148-149, 151-152, 182-183
	(5) Cosmology	129
III.	Geography and Geodesy:	
	(1) Mathematical Geography	20-21, 27-33, 114
	(2) Cartography	19, 22-23, 112
	(3) Geodesy	24-25, 109, 136
	(4) Climatology	26, 68, 131, 154
IV.	Physics and Natural Sciences	
	(1) Densimetry	63
	(2) Mineralogy	156-157, 159
	(3) Medicinal Botany	158
V.	Chronology and History	
	(1) Origins	55
	(2) Chronology	53-54, 103, 105
	(3) Annals	150, 160-162
VI.	Literature and Philosophy	
	(1) Stories	80-85, 87-92
	(2) Bibliography	168
	(3) Grammar	138
	(4) Poetry	86, 163-166
	(5) Philosophy	130, 133-134, 153, 170, 172-173
VII.	Customs and Religion	
	Miscellaneous Works	93-98, 174

In regards to classification of Biruni's works, he must be seen as the best judge of his own research agenda. We believe that his own list of books and classification in the *Fihrist* still is the most accurate. But he provided the names of only 138 books and his classification is nearly one thousand years old. It was done at a time when boundaries between different disciplines were not clearly drawn. In his classification, Table 1, five out of sixteen categories are entitled "Miscellaneous." In fact, for all of the categories except one, Customs and Religion, explanatory keywords such as the names of modern disciplines are added in parentheses.

Khan's classification of Biruni's works, given in Table 2, divides the 183 books into 7 main categories with 26 sub-categories. He too could not avoid having a sub-category called "Miscellaneous Works." We could assume that Biruni wrote some books that still cannot be simply classified under any one of the 26 headings. Khan's classification, of course, makes more sense. He uses the names of many modern disciplines including Trigonometry, Cartography, Geodesy, and Climatology. Obviously, these names were not available during Biruni's life. To find desired topics among Biruni's works, however, we recommend that the readers consult both tables.

As mentioned before, Biruni's research themes topped every discipline in the annals of human intellectual history. His research agenda encompassed all areas of research from Anthropology to Zoology and simply left no area of research untouched. According to Durant (1950), Biruni was a "philosopher, historian, traveler, geographer, linguist, mathematician, astronomer, poet, and physicist-and doing major and original work in all these fields..." (p. 243). Biruni's early method of writing was extremely simple but slowly developed into more complicated subjects to a degree that his students at advanced levels began to complain.

In his very first book, *Al-Masa'il al-Mufida wa'l Jawabat al-Sadida*, translated as "Useful Questions and Right Answers", Biruni's approach is very simple. As can be seen from the title, he dealt with an important subject, Astronomical Tables, in a "Question-Answer" format. In this beautiful title, the two Arabic words "*Mufida* and *Sadida*" rhyme. This is the case for most of the titles of his books. The words in the titles are chosen so thoughtfully that they become exquisite pieces of poetry by themselves.

The Persian version of entry number 73, *Kitab al-tafhim li-awa'il sina'at al-tanjim*, was one of Biruni's books that was available to us. Fortunately, we know a little more about this book. We know that Biruni wrote this book for Princess Raihanah and translated it into Persian from Arabic for her, because she was unable to understand Arabic. The keywords in this title include: *Tafhim*, *Awa'l*, and *Tanjim*, that literally mean: "Understanding, Beginning, and Astronomy." These keywords indicate that Biruni's purpose was to write a book to make this science accessible to beginners who just want to understand the basics. Also, notice how the two Arabic words of *Tafhim* and *Tanjim* beautifully compose a rhyme. Unfortunately, R. Ramsay Wright translated the Arabic word *Tanjim* into English as Astrology instead of Astronomy in 1934, and Khan repeats this mistake. Biruni himself expressed his disdain for alchemy and astrology. The book contains no subject associated with the modern understanding of astrology. In Arabic and Persian, the word *Tanjim* means "studying stars," not fortune telling. In fact, Biruni's very first sentence in the book defines "Astronomy."

Al-Tafhim, as it is most often called for short, is in itself an enormous achievement for a scholar in the Medieval Ages. The Persian version of this book that was available to us was edited by the late Professor Humai of the University of Tehran, consisting of a general brief introduction, a biography of Biruni, a long introduction to *Al-Tafhim*, a correction section, and an index in 924 pages. In this book too, Biruni's method is to ask questions and provide answers in the simplest possible way. For example, some of the questions in the first part of the book ask, What is geometry? What is a point? What is an angle? On the other hand, according to Humai " although this book was written for a young Iranian Princess about a thousand years ago, yet it contains subjects and problems that still are not fully understood in our time" (p. 23).

In *Al-Tafhim*, Biruni not only has laid an outstanding foundation for the science of astronomy but this book is also an important piece of literary work. This book became a model for innovative scientific terminology in Arabic and their beautiful parallel translations in Persian. Many of these standardized keywords both in Arabic and Persian were used by future generations of scholars.

Since the eighth century many Muslim leaders and scholars have been interested in expansion of their religion and trade in a large empire and with others. With the growth of the empire and the increase in distance from its hearth, came an effort to learn about different places in the world. First of all, Muslims needed to know the correct direction to face the holy city of Mecca for their daily prayers. Thus, many Muslim leaders and scholars, including Biruni, were concerned about correct and scientific determination of location and direction.

The general direction of daily prayers towards Mecca is called the Qibla. The titles of six of Biruni's books include this word. Abdali (1997) argues that "the most important Muslim contribution to mathematical geography was a treatise [our entry number 19] by the eleventh-century scientist al-Biruni. His purpose was to determine for his patron the qibla at Ghazna, in what is now Afghanistan, a goal which he achieved most admirably" (p. 4).

Two of Biruni's books, numbers 19 and 112, specifically deal with determination of location. His calculations of latitudes and longitudes, coordinate systems, for many cities in the world are still valid, not very different from their modern values. Entry number 19, *Kitab Tahdid nihayat al-amakin li tashih masafat al-masakin*, has already been translated into English. It is called *Determination of the Co-ordinates of Positions for the Correction of Distances between Cities.*

Biruni must be given credit for coining the word "Azimuth." Azimuth is defined as the angular distance along the horizon with zero degrees corresponding to North and increasing in a clockwise fashion. The word ***Al-Sumut***, a plural form of the Arabic word ***Al-Samt***, which means "Directions", appeared in three of the titles of his books, entry numbers 16, 114, and 119. Actually, entry 114, *A Book on Azimuths*, should be considered the first book devoted entirely to this subject. Many dictionaries acknowledge the origin of the word "Azimuth" as being Arabic. Indeed, he deserves to be known as the founding father of the science of Geodesy. Almost all of his research agenda dealt with the practical application of science to real-world situations.

While carefully looking at 183 book titles and writing brief annotations, one also observes different stages of Biruni's intellectual development. When he was very young he had to find a niche for himself among many scholars of his time, a quest that was no easier then than now. Biruni's plan to make himself acceptable to the society of scholars was interesting. Since childhood, Biruni was very serious about reading books. And, he had always strongly opposed any half truths, misleading or erroneous information. He was also bitterly against unfounded beliefs, superstition, myth, and metaphor. His earliest accomplishment was to detect many astronomical miscalculations. Thus, his earlier books deal mostly with rejection of many misleading writings that were accepted as facts in his time. Finding the truth was always Biruni's major goal in life. After rejecting those myths, he presented his own findings in the form of the most original scientific books and writings. It is no wonder that eleven of his earlier books contain the word "correction" in their titles.

When Biruni was about twenty years old, he was recognized as a respected and serious scholar by the royal courts of several kingdoms and by his colleagues. After he wrote his first major book, *Al-Athar al-Baqiya an al-Qurun al-Khaliya* , translated as the *Chronology of Ancient Nations*, number 105 in our list, he felt comfortable with his ability and was widely accepted as a young scholar of great standing. After this stage he knew what he was doing. His goal was to find and document more truths, and he was not to be challenged.

In *Chronology*, Biruni provided information on all approaches to reckoning of time and religious holidays in the Middle Ages. It is still a major reliable source for historians of diverse religions. This book is one of few known reliable sources of information on ancient Persian history. It is believed that Jews have rediscovered their calendar and holidays from Biruni's *Chronology*.

Biruni was a gifted inventor and had an enormous appetite for devising tools for use in his scholarly fields. He built a very large globe and modified and constantly improved astrolabes of many different types. An astrolabe is an instrument for observing the celestial bodies. It is a very old astronomical computer for solving problems relating to time and the position of the Sun and stars in the sky. With a long history, the astrolabes are believed to be the oldest and the most sophisticated scientific instrument in the world. Seven of Biruni's book titles contain the world "Astrolabe." These entries include numbers 46, 47, 49, 50, 119, 169, and 177.

In regard to his ability to write in different languages or translate books, Biruni has no parallel in the annals of scientific history. As mentioned before, Biruni had extraordinary enthusiasm for obtaining objective knowledge. This noble goal encouraged his hunger to learn foreign languages. He obliged himself to go directly to the original sources and ethnic groups. He learned at least eight different languages and utilized them in his research.

Most of the titles of the books in the last section of the following list are translations. A large number of books that he translated were from Sanskrit into Arabic or vice versa. We found a total of 26 books out of 183 that were translations. Of these only one book, entry number 77, does not contain the word "Translation" in its title. Actually, the titles of the other 25 books begin with "Translation of…." They include entry numbers 6, 40, 80-92, 97-98, 159, 161, 174-177, 181, and 183. We know that Biruni was very meticulous in giving credit to his references. In a sense, he acknowledged the fact that a book was not originally written by him by using "Translation of…" in the opening of the title of the book. Also, this recognizable indication can be used in classification of his books. With these many translated books, Biruni's intense interest and ability in learning and using other languages must have been very fruitful.

Biruni is considered to be the first, the best, and the most honest scholar who presented India and Indian civilization in the world. His books about India and Indian civilization are still among the best available references. We found the word "India" in 18 of the titles of his books, the most important being number 93, *Kitab fi Tahqiq ma li'l-Hind*, known as *Al-Biruni's India*. Biruni is the first and the only scholar who argued that Indian advancement of knowledge was not far behind that of the Greeks. This book still is an important reference on medieval India even for the Hindu scholars themselves. In 1968, Mr. Ghayasuddin wrote his doctoral dissertation on Biruni's writings about India. It was entitled *A Critical Analysis of the Writings of Abu'l Raihan Muhammad al-Biruni, Pertaining to India.*

The books that were written late in his life show a more mature and confident scholar. His *Canon of Masudi*, our number 104, is believed to be one of the most complete sources of knowledge, especially advanced and more sophisticated astronomy. His books on gem stones, medicinal drugs, history of his homeland, and a bibliography, *al-Fihrist*, numbers 156, 158, 160, 168, are among the first and are still among the best available.

Biruni was not a grim scholar inundated in scientific reasoning. He appreciated and enjoyed fine arts and poetry. He wrote or translated poetry. Indeed, four of the titles of his books, numbers 86, 163, 164, and 165 include the word "poetry". In addition, another four of his books, numbers 80, 81, 84, and 85 are about love stories. One of his books deals with calligraphy and another one about etiquette. At the same time, records suggest that he was a practical joker.

Before providing the list of Biruni's books, we would like to offer a note on the Arabic-Persian names of the books. The titles of Biruni's 183 books were first typed using software called *Vazhe Negar*. Then, these titles were inserted in their appropriate places, one by one, in our *Microsoft Word* file. At this stage, the *Word* file did not recognize these names. Fortunately, the *Word* file did communicate with *Vazhe Negar* for using Arabic-Persian fonts. Unfortunately, the *Word* file was unable to translate and keep the accents of the characters in their correct locations.

These accents, essential in reading and assuming correct meaning of the words in Arabic, were shown as empty spaces, and single words were divided, thus creating many errors. Thus, we had to delete those spaces. When read with the accompanying title transliterations our Arabic titles become meaningful. However, for those readers who are used to seeing and reading Arabic words with accents, we have furnished a complete list of the books in Arabic with accents at the end of this book in Appendix A.

Sadly enough, of the 183 books, few have survived. From a dozen or so books that have been found, we can choose and recommend eight of Biruni's masterpieces, including numbers: 19, 93, 104, 105, 156, 158, 160, and 168. It is worth the time to study carefully the names of Biruni's books in the following list. These are among the books that changed the history of human thinking.

Biruni's 183 Books in Arabic, Persian, Transliteration, English and a Brief Annotation:

۱ـ المسائل المفيده و الجوابات السديده.
(پرسشهای سودمند و جوابهای درست)

1. *Al-Masa'il al-Mufida wa'l Jawabat al-Sadida.* (Useful Questions and Right Answers).

According to Khan this book had 1250 folios. Yet, Azkayi's number of folios is only 250. It was written on the astronomical tables of Muhammad Ibn Musa Khwarazmi (780-850 C. E.), a famous Muslim astronomer-mathematician.

۲ـ ابطال البهتان بايراد البرهان على عمل الخوارزمى فى زيجه.
(باطل كردن تهمت ها با آوردن دليل براى كارهاى خوارزمى در زيج او)

2. *Ibtal al-Buhtan bi-'irad al-Burhan ala amal al-Khwarizmi fi Zijihi.* (The Annulment of Lie by Providing the Proof on the Work of Khwarizmi Concerning His Astronomical Tables).

This book is a refutation of the physician Abu Talha's treatise on the astronomical tables of Khwarazmi. This was written in 360 folios.

۳ـ كتاب الوساطه بينهما.
(كتاب ميانجيگرى بين آنها، خوارزمى و اهوازى)

3. *Kitab al-Wasata baynahuma.* (The Mediation Between the Two).

This book was written in 600 folios. Biruni found a book by Abu'l Hasan al-Ahwazi in which the author did not do justice to the astronomical tables of Khwarazmi. Al-Ahwazi translated a book from Sanscrit into Arabic before the year 1000 C. E. Biruni attempted to correct this injustice.

۴ـ تکمیل زیج حبش بالعلل و تهذیب أعماله من الزلل.
(تکمیل زیج حبش با بیان علتها و پیراستن کارهای او از لغزشها)

4. *Takmil Zij Habash bi'l-ilal wa tahdhib amalihi min al-Zalal.* (Improvement of the Tables of Habash with Causes and Correction of His Works from Errors).

This book was written on the astronomical tables of Habash in 250 folios. In fact, this work was an improvement of the previous calculations by Ahmad Ibn Abdallah Habash (died 870 C. E.). Habash's nickname was "Hasib" which in Arabic means "One Who Can Calculate." He calculated tables of sines, tangents, and standard astronomical functions. Biruni corrected his mistakes and updated the knowledge on the subject of the Zij, astronomical tables.

۵ـ جوامع الموجود لخواطر الهنود فی حساب التنجیم.
(مجموعهٔ آراء خاطرهای هندیان در بارهٔ محاسبات احکام نجوم)

5. *Jawami al-Maujud li-Khawatir al-Hunud fi Hisab al-Tanjim.* (Collection of the Ideas of the Indians on the Astronomical Calculations).

A book on *Sindhind* in 550 folios partly completed. The word *Sindhind* or *Siddhanta* indicates the astronomical works of Indian astronomers.

۶ـ تهذیب زیج الارکند.
(پیراستن زیج ارکند)

6. *Tahdhib Zij al-Arkand.* (Corrections of the Astronomical Tables of al-Arkand).

This book is a presentation in Arabic of the astronomical tables of *Arkand* which are made understandable by translation of the Indian technical terms. *Arkand* is a corruption of the Indian *Ahargana* meaning "Total Days." *Khandakhadyaka* of Brahmagupta's work on Indian astronomy was already translated into Arabic long before Biruni's time. But, the translation was not very accurate and contained many errors and omissions. Thus, Biruni produced a superior translation based on his understanding of the subject. The number of folios is not known.

۷ـ کتاب مقالید علم الهیئت و ما یحدث فی بسیط الکره.
(کتاب گنجینه های علم هیئت و گفتگو در بارهٔ کره)

7. *Kitab Maqalid ilm al-hay'a ma yahdithu fi basit al-Kurra*. (Keys of Astronomy on What Takes Place on the Surface of the Sphere).

This book deals with the science of astronomy with specific reference to the sphere. It was written for Marzban Ibn Rustam, independent governor of Tabaristan province, southeast of the Caspian Sea, in Iran. Here Biruni has shown methods of determining the degrees of the Zodiac and also certain stars in the middle of the sky. The work was in 155 folios. Leiden Manuscript No. 1066.

۸ـ خیال الکسوفین عند الهند.
(تصور کسوف و خسوف در نزد هندیان)

8. *Khayal al-Kusufayn ind al-Hind.* (The Image of the Two Eclipses by the Indians).

This book was written on the image of two equal circles that unite into one, representing the eclipse, whose nature was well understood by Indians. The number of folios is not known.

۹ـ فی امر الممتحن و تبصیر ابن کیسوم المفتتن.
(در بارهٔ زیج ممتحن و بینا کردن ابن کیسوم برآشفتهٔ شوریده)

9. *Fi amr al-Mumtahan wa-tabsir Ibn Kaysum al-Muftatan*. (On the Examined Tables and the Faulty Explanation of Ibn Kaysum).

Biruni exposed Ibn Kaysum's shallow knowledge of astronomy. This book deals with the *Zij al-Mumtahan*, "Examined Tables" that were prepared by the astronomers of the Caliph Mamun in Baghdad. This book was written in 100 folios.

۱۰ـ اختلاف الاقاویل لاستخراج التحاویل.
(اختلاف اقوال در بارهٔ استخراج تحویل، سالها و بروج، نقل و انتقال ستارگان در برجها)

10. *Ikhtilaf al-aqawil li-istikhraj al-tahawil*. (Different Sayings on Calculating the Revolutions).

This work is a treatise written to answer questions of a scholar on the "revolutions of the years." The work was in 30 folios.

۱۱ــ مقاله فی التحلیل و التقطیع للتعدیل.
(مقاله ای در مورد قطع کردن و تعدیل خورشید)

11. *Maqala fil-tahlil wa 'l-taqti ' li 'l-tadil.* (Discourse on the Resolution and Analysis of the Equation).

This treatise was addressed to a late scholar who was unsure of his calculations on the equation tables of the Sun. Biruni determined that, indeed, he had not followed a good method of resolution. The work was in 30 folios.

۱۲ــ فی تهذیب الطرق المحتاج الیها فی استخراج هیئت الفلک عند المواليد و تحاويل السنين و غيرها من الاوقات.
(پیراستن روشها که در استخراج هیئت فلک از لحاظ زایچه ها و تحویل سالها و جز آنها از بابت اوقات مورد نیازند)

12. *Fi tahdhib al-turuq al-Muhtaj ilayha fi istikhraj hay'at al-falak ind al-Mawalid wa Tahawil al-sinin wa-ghayriha min al-auqat.* (Improvements of the Methods Used in the Calculation of the Configuration of the Years and Other [Aspects] Relating to Time).

This is a treatise in 60 folios. It deals with astrological subjects and especially with horoscopes.

۱۳ــ مفتاح علم الهیئت.
(کلید علم هیئت)

13. *Miftah ilm al-Hay'a.* (Key to Astronomy).

This book was written for Qasim al-Amiri. It deals with principles of astronomy. It was cited in Biruni's *India.* Biruni himself translated this work into Sanskrit. The number of folios in this work was 30.

۱٤ــ تهذیب فصول الفرغانی.
(پاک کردن فصل های کتاب فرغانی)

14. *Tahdhib fusul al-Farghani.* (Correction of the Chapters of al-Farghani's Book on Astronomy).

This book was written for Abul Hasan Musafir on Farghani's astronomy work. The work was in 200 folios.

۱۵ـ کتاب فی افراد المقال فی امر الاظلال.
(جدا ساختن مبحث در بارهٔ سایه ها)

15. *Kitab fi ifrad al-Maqal fi 'amr al-azlal.* (The Exhaustive Treatise on Shadows).

This book, also written for Abul Hasan Musafir, deals with the measurement of straight lines called "Zill" or "shadows", denoting tangents or cotangents, in the science of surveying, and several other related problems. The book was in 200 folios. This book was translated by E. S. Kennedy as *The Exhaustive Treatise on Shadow*. Aleppo, Syria: Institute for History of Arabic Sciences, 1967, with commentary, in 2 volumes. Manuscript: Bankipore 2468/36=Patna 2,335,2519/34.

۱۶ـ کتاب فی استعمال دوائر السموت لاستخراج مراکز البیوت.
(کاربردن دایره های سمتها برای استخراج مرکزهای خانه ها)

16. *Kitab fi istimal dawa'ir al-sumut li-istikhraj marakiz al-buyut.* (Employment of the Circles of Azimuth for the Calculation of the Centers of the Houses).

This was another book written for Aul Hasan Musafir. Here, Biruni explains an equation, *Taswira*, on the significance of this astrological expression according to Muslim sources. It was written in more than 100 folios.

۱۷ـ مقاله فی طالع قبه الارض و حالات الثوابت ذوات العروض.
(مقاله ای در مورد طالع گنبد زمین و حالتهای ستارگان ثابت عرضدار)

17. *Maqala fi tali qubbat al-ard wa halat al-thawabit dhawat al-urud.* (On the Rise of the Cupola of the Earth and the Position of the Fixed Stars with Latitudes).

This treatise describes the center of the Earth and those stars that are situated north of the Equator. This treatise was written for an astrologer in the city of Gorgan, southeast of the Caspian Sea. It was in 100 folios.

۱۸ـ مقاله صغیره فی اعتبار مقدار اللیل و النهار فی جمع الارض لتعریف کون السنه یوماً تحت القطب بغیر تشکیل.
(مقاله ای کوتاه در مورد تامل مدت شب و روز در سراسر زمین برای شناساندن وجود سال یک روزه در زیر قطب بدون صورت بستگی)

18. *Maqala saghira fi itibar miqdar al-layl wa'l nahar fi jami al-ard li-tarif kaun al-sanat yawman taht al-qutub beghaya tashkil.* (A Small Discourse on the Consideration of the Duration of the Night and the Day throughout the Earth, Defining that at the Pole, a Year Is Equal to a Day.)

This is a small tract without figures. Biruni believed that in the polar region one day is like a complete year, half of the year is day and the other half, a night. The number of folios is not given in our sources.

۱۹ـ کتاب تحدید نهایات الاماکن لتصحیح مسافات المساکن.
(کتاب تعیین حدهای نهایی مکانها برای تعیین فاصله های جایگاه ها)

19. *Kitab Tahdid nihayat al-amakin li tashih masafat al-masakin*. (The Determination of the Co-ordinates of Positions for the Correction of Distances between Cities).

This book was in 100 folios. It has a strong relationship with *Qanun al-Masudi*, number 104, below. Khujandi's mistakes, due to the weight of his sextant, are reported in this book (see number 169 below). Here, Biruni reported his measurements of the radius, diameter, and circumference of the Earth that are very similar to their modern values. This book was translated into English by Jamil Ali and was called the *Determination of the Co-Ordinates of Positions for the Correction of Distances Between Cities*. Beirut: American University of Beirut, Centennial Publications, 1967, xviii, 278 pp. Manuscript: Sultan Fatih 3386.

۲۰ـ کتاب تهذیب الاقوال فی تصحیح العروض و الاطوال.
(کتاب پیراستن گفته ها در بارهٔ تصحیح عرضها و طولهای جغرافیایی)

20. *Kitab Tahdhib al-Aqwal fi tashih al-Urud wa-l-atwal*. (Improvement of Statements on Correction of the Latitudes and Longitudes).

This book is also somewhat related to *Qanun al-Masudi*, number 104, below. This book was in 200 folios.

۲۱ـ کتاب تصحیف المنقول من العرض و الطول.
(کتاب خطا خوانی و غلط نویسی در مورد عرضها و طولهای جغرافیایی)

21. *Kitab tashif al-manqul min al-ard wal-tul*. (Corrections of the Traditional Errors Relating to the Latitude and Longitude).

This is another of Biruni's books related to *Qanun al-Masudi* (number 104 below). This book was written in 40 folios.

۲۲_ مقاله فی تصحیح الطول و العرض لمساکن المعمور من الارض.
(مقاله ای در مورد تصحیح طول و عرض جغرافیایی جایگاهای آبادان زمین)

22. *Maqala fi tashih al-tul walard li-masakin al-mamur minal-ard.* (Corrections of the Longitudes and the Latitudes of the Cities of the Inhabited Part of the World).

This work was combined with number 23 and was related to *Qanun al-Masudi*, (number 104 below). This work consists of 20 folios.

۲۳_ مقاله فی تعیین البلد من العرض و الطول.
(مقاله ای در مورد مشخص کردن شهر از لحاظ طول و عرض جغرافیایی)

23. *Maqala fi tayin al-balad min al-ard wa 'l-tul.* (Fixation of the Longitude and the Latitude of a City).

This work deals with the placing on a map of a city whose latitude and longitude are known. It consists of 20 folios.

۲٤_ مقاله فی استخراج قدر الارض برصد انحطاط الافق عن قلال الجبال.
(مقاله ای در مورد محاسبهٔ اندازهٔ محیط زمین با رصد کردن انحطاط افق از بالای قلهٔ کوهها)

24. *Maqala fi istikhraj qadr al-ard bi-rasd inhitat al-ufuq an qulal al-jibal.* (Discourses on Obtaining the Value of the Earth by Measuring the [Angle] of Declination of the Horizon from the Summits of Mountains).

The Berlin Manuscript No. 5794 (Ahlwardt's catalog, 5) contains only a brief extract of the original work. This was a treatise in 60 folios.

۲۵_ فی غروب الشمس عند منارات الاسکندریه.
(در مورد غروب آفتاب بر روی منارهٔ شهر اسکندریه)

25. *Fi Ghurub al-Shams ind manarat Iskandriya.* (On the Setting of the Sun at the Lighthouse of Alexandria).

This book was written on the famous lighthouse at Alexandria, Egypt. This work was written in 40 folios.

۲٦ـ فی اختلاف الواقع فی تقاسیم الاقالیم.
(در مورد اختلاف رخداده در تقسیمات اقیمها)

26. *Fi 'likhtilaf al-Waqi fi taqasim al-aqalim*. (On the Various Opinions on the Division of the Clime).

This work , in 20 folios, was written on different types of climates of the known world.

۲۷ـ فی اختلاف ذوی الفضل فی استخراج العروض و المیل.
(در مورد اختلاف فضلمندان در استخراج، محاسبهٔ، عرض ها و میل ها)

27. *Fi ikhtilaf dhawi l-fadl fi istikhraj al-urud wa'l mil.* (Differences of Opinion Among the Learned on Obtaining the Latitude and the [angles] of Declination).

This work was about the coordinate systems. The number of folios is not known.

۲۸ـ کتاب الاجوبه و الاسئله لتصحیح سمت القبله.
(کتاب پرسشها و پاسخها از برای تصحیح سمت قبله)

28. *Kitab al-Ajwiba wa-l-as'ila li-tashih simt al-Qibla*. (Questions and Answers Relating to the Correction of the Direction of the Qibla).

This work discusses methods that were utilized for determination of the direction of the Qibla or Kiblah, the direction to which Muslims turn in praying. This work was in 30 folios.

۲۹ـ ایضاح الادله علی کیفیت سمت القبله.
(توضیح دلایلی که بر چگونگی سمت قبله آورده شده)

29. *Idah al-adilla ala kayfiyat simt al-Qibla*. (Exposition of the Proofs on the Nature of the Direction of the Qibla).

This work is an explanation of the proofs on the characteristics of the direction of the Qibla. It was in 25 folios.

۳۰ـ تهذیب شروط العمل لتصحیح سموت القبله.
(پیراستن شرطهای عمل برای تصحیح سمتهای قبله)

30. *Tahdhib shurut alamal li-tashih sumut al-Qibal.* (Rectification of Conditions Operative in the Work of Correcting the Directions of Qiblas).

This work lays down basic conditions that are essential to determine the direction of the Qibla. It was written in 40 folios.

۳۱ـ فی تقویم القبله فی بست بتصحیح طولها و عرضها.
(در مورد درست کردن جهت قبله در شهر بست با تصحیح طول و عرض آن)

31. *Fi taqwim al-qibla fi-Bust fi-tashih tuliha wa-ardiha.* (Determination of the Qibla at Bust by Correcting Its Latitude and Longitude).

This work was written on determining the direction of Qibla at the city of Bust, in the Sijistan district on the Hehmand River, to the west of the city of Qandahar in Afghanistan today. It was in 15 folios.

۳۲ـ فی الانبعاث لتصحیح القبله.
(اقدام برای تصحیح جهت قبله)

32. *Fi'l-inbiath li-tashih al-Qibla.* (The Process for the Correction of the Qibla).

This work is also about determining the correct direction for Qibla. It was written in 45 folios.

۳۳ـ تلافی عوارض الزله فی کتاب دلائل القبله.
(دریافت و جبران لغزشهای رخداده در کتاب رهنمایی های قبله)

33. *Talafi awarid al-zalla fi Kitab dala'il al-Qibla.* (Correction of the Mistakes in the Book *Dalail al-Qibla).*

This book is an attempt to correct mistakes in a book called *Dala'il al-Qibla* or *Directions of Qibla.* The number of folios is not given.

۳٤ــ تذکره فی الحساب و العد بارقام السند و الهند.
(تذکره ای در بارهٔ حساب و شمارش با رقمهای سند و هند)

34. *Tadhkira f'il-hisab wall-add bi-arqam al-Sind wa'l-Hind.* (Calculation and Enumeration According to the Numerals of Sind-Hind).

This is a memoir that is related to the Indian numerals. It was in 30 folios.

۳٥ــ فی استخراج الکعاب و اضلاع ماورائه من مراتب الحساب.
(در مورد استخراج کعب ها و جذرهای آنچه فراسوی حساب است)

35. *Fi istikhraj al-kiab wa-adla mawara'ahu min maratib al-hisab.* (Obtaining the Cubes and Square Roots and Other Rules of Arithmetic).

The rules of arithmetic, especially cubes, are described in this work. This work was written in 100 folios.

۳٦ــ کیفیت رسوم الهند فی تعلم الحساب.
(در مورد چگونگی روشهای هندیان در آموختن حساب)

36. *Kayfiyat rusum al-Hind fi taallum-al-hisab.* (Indian Rules for Learning Arithmetic).

This work describes the Indian rules of learning arithmetic. The number of folios is not known.

۳۷ــ فی ان رای العرب فی مراتب العدد اصوب من رای الهند فیها.
(در مورد اینک نظر تازیان در مراتب عدد درستر از هندیان است)

37. *Fi anna ra'y al-Arab fi maratib al-'adad aswab min ra'y al-Hind fiha.* (The Superiority of the Arab Opinion to that of the Indians Relating to the Rule of Numbers).

This work claims that the knowledge of the Muslims about numeral systems was far superior to that of the Indians. It was in 15 folios.

۳۸ــ فی رشیکات الهند.
(در مورد تناسب های هندی)

38. *Fi Rashikat al-Hind.* (On the Indian Rules of the Three).

This work deals with the method of calculating specific gravity. It was written in 15 folios. Manuscript: Bankipore 2468/37=Patna 2,335,2519/35—and Indian Office Library 1043/1 (incomplete).

۳۹ــ فی سکلب الاعداد.
(در مورد سکلب عددها)

39. *Fi (Saklab?) al-adad.* (On the "Saklab" of the Numbers).

The significance of the word "Saklab" cannot be ascertained but may deal with a method of calculation. According to Kahn, half of the work is in 30 folios, but Azkayi's information indicates 40 folios.

٤٠ــ ترجمه ما فی براهم سدهاند من طرق الحساب.
(ترجمهٔ آنچه در "برهمگمپت سدهانت" مربوط است به راههای محاسبه)

40. *Tarjamat ma fi Brahma Siddhanta min turuq al-hisab.* (Translation of the Methods of Calculations as Found in Brahmasiddhanta).

Biruni made important contributions to Indian research by translating Indian documents from Sanskrit into Arabic. This work is a translation of *Brahmasiddhanta.* This work consists of 40 folios.

٤١ــ منصوبات الضرب.
(شیوه های ضرب)

41. *Mansubat al-darb.* (Methods of Multiplication).

This work deals with different intricate methods of multiplication and its attributes. The number of folios is not known.

٤٢ــ تجرید الشعاعات و الانوار عن الفضائح المدونه فی لاسفار.
(پیراستن شعاعات و انوار از گفتارهای رسوا و ناروایيها که در کتابها تدوین گردیده)

42. *Tajrid al-Shuaat wa'l-anwar an al-fada'ih al-mudawwana fi'l asfar.* (Clearing the Recorded Ignominies Pertaining to Rays and Light in Passing).

This book attempts to correct misconceptions related to rays of light; on the astrological notion of rays and their projections. It contains 55 folios.

٤٣ـ مقاله فی تحصیل الشعاعات بابعاد الطرق عن الساعات.
(مقاله ای در مورد حاصل کردن شعاعها با دورترین راه از ساعتها)

43. *Maqala fi tahsil al*-shuaat bi-ab ad al-turuq an al-saat. (Discourse on Obtaining the Rays by the Most Remote Methods of the Hours).

This is a treatise in 10 folios. While the title of this work is not clear in its intended meaning, it has been described as a work dealing with obtaining the condition of rays by using time methods.

٤٤ـ فی مطرح الشعاع الثابتاً علی تغییر البقاع.
(در مورد جای پرتو افکندن بطور ثابت بنا بر تغییر مکانها)

44. *Fi Matrah al-shua thabitan ala taghayyir al-biqa.* (On the Place of Ray's Occurrence Which Remains Fixed In Spite of the Change in the Region).

This is another treatise on rays in 15 folios. Biruni himself listed this work under the category of astrological notions.

٤٥ـ تمهید المستقر لتحقیق معنی المر.
(ترتیب قرارگاه برای تحقیق معنای گذار)

45. *Tamhid al-mustaqarr li-tahqiq mana al-mamarr*. (Smoothing the Basis for an Investigation of the Meaning of Transits).

This work deals with the subject of shadow-lengths. The meaning of the word *Mamar* is a place of passing or transit. Here it is used in the context of rays. This work is in 60 folios. This book was translated into English by M. Saffouri and A. Ifran, with commentary by E. S. Kennedy, as *Al-Biruni on Transits*. Beirut: American University of Beirut, 1959, ix, 201 pp. Manuscript: Bankipore 2568/38=Patna 2,335,2519/36.

٤٦ـ کتاب فی استیعاب الوجوه الممکنه فی صنعت الاصطرلاب.
کتاب اخذ و برسی کامل روشهای ممکن در فن اسطرلاب)

46. *Kitab fi istiab al-wujuh al-mumkina fi sanat al-Asturlab*. (Exhaustive Study of the Possible Methods for the Construction of Astrolabe).

This is a book in 80 folios. The work is very rich in content. It gives a description of the apparatus used to divide circles and straight lines, and identifies different forms of astrolabes. Manuscript: Aya Sofia No. 2576.—Ashir No. 572.—Berlin No. 5795.—Paris No. 2498.—Bodleian, 1, 1037/3, 16 pp. (only dealing with the universal Astrolabe). British Museum Oriental 5593 (DL 38).—Jarullah, No. 1451.—Leiden No. 1066.—Rampur, 1, 425.—Topkapi Serai No. 3505/7.—Tehran, 2, 150.— Zaytuna (Tunis) No. 5593 & 5540.

٤٧_ فی تسهیل التصحیح الاصطرلابی و العمل بمرکباته من الشمالی و الجنوبی.
(در مورد آسان کردن تصحیح اسطرلابی و کار کردن با ترکیبات شمالی و جنوبی آنها)

47. *Fi tashil al-tashih al-usturlabi wa'l-amal bi-murakkabatihi min al-shumali wa'l-junubi.* (Easy Method for the Correction of the Astrolabes and the Use of Its Northern and Southern Compound Devices).

This book deals with correcting astrolabes that projected part of the eclipse on the Equator going towards the North Pole and another going to the South Pole. This work has only 10 folios.

٤٨_ فی تسطیح الصور و تبطیح الکور.
(در مورد تصویر صورتهای فلکی و ترسیم نواحی بر سطح)

48. *Fi tastih al-suwar wa-tabtih al-kuwar.* (Plane Projection and Spreading Out of the Constellations and the Countries).

This work is also in 10 folios. It deals with plane projection and was classified under astrological notations.

٤٩_ فیما اخرج ما فی قوه الاصطرلاب الی الفعل.
(آنچه در مورد اسطرلاب از قوه به فعل حاصل شده است)

49. *Fima ukhrija ma fi quwwat al-usturlab ila'l-fil.* (On the Power of the Astrolabe in Its Action).

This work deals with the problems in which the astrolabe can be used and explains different methods of using them. It is in 30 folios.

۵۰_ فی استعمال الاصطرلاب الکروی.
(در مورد کاربرد اسطرلاب کروی)

50. *Fi 'istimal al-usturlab al-kuravi*. (On the Use of the Spherical Astrolabe).

This work is written on how to use the spherical astrolabe. It has 10 folios. Biruni listed this work under instruments.

۵۱_ مقاله فی تعبیر المیزان لتقدیر الزمان.
(مقاله ای در مورد سنجش ترازو برای اندازه گیری زمانها)

51. *Maqala fi tabir al-mizan li-taqdir al-azman*. (Explanation of the Balance for the Measurement of Time).

This work in 15 folios is a treatise on time and movements. It describes the "balance" by which time could be determined.

۵۲_ فی تحصیل الآن من الزمان عند الهند.
(در مورد دریافت کنون از زمان در نزد هندیان)

52. *Fi tahsil al-'an min al-zaman ind al-Hind*. (Ascertainment of the Present Moment of Time According to the Indians).

This work deals with the Indian method of measuring time to the minute and second. This book was written in 100 folios.

۵۳_ تذکره فی الارشاد الی صوم النصاری و الاعیاد.
(تذکره ای در مورد رهنمونی به سوی روزهٔ مسیحیان و عیدهای آنان)

53. *Tadhkira fi 'lirshad ila sawm al-Nasara wa 'l-ayad*. (An Account of the Guide to the Feasts and Festivals of the Christians).

This is a memoir in 20 folios. This work describes Christian holidays and feasts that were common a thousand years ago.

۵۴_ فی الاعتذار عما سبق لی فی تاریخ الاسکندر
(در مورد پوزش از آنچه پیشتر در تاریخ اسکندر آورده ام)

54. *Fi l-itidhar amma sabaqa li fi tar'ikh al-Iskandar*. (Apology [for the Mistakes] in the History of Alexander).

This work corrects the mistakes that were made earlier in a book on the history of Alexander. It consisted of 10 folios.

۵۵ـ فی تکمیل حکایات عبدالملک الطبیب البستی فی مبدا العالم و انتهائه.
(در مورد کامل کردن بازگوییهای عبدالملک پزشک بستی در بارهٔ آغاز و انجام جهان)

55. *Fi takmil hikayat Abd al-Malik at-tabib al-Busti fi mabda' al-alam wa intiha'ihi.* (Completion of the Accounts of the Physician Abd al-Malik al-Busti on the Origins and the End of the World).

This book was written in 100 folios. A physician, Abd al-Malik Busti, gave an account of the origins and end of the world. He was from the city of Bust, now in Afghanistan. Biruni completed these accounts based on his own knowledge and added more information on the subject. He narrated beliefs of different nations on this topic.

۵۶ـ مقاله فی دلالت الآثار العلویه علی الاحداث السفلیه.
(دلالت پدیده های آسمانی بر رخداده های زمینی)

56. *Maqala fi dalalat al-'athar al-ulwiya ala 'l-ahdath al-sufaliya.* (Discourse on the Influence of the Celestial Phenomena on Earthly Events).

This treatise deals with comets of different types, some with tails, some with "hair", and some with stars. This treatise is in 30 folios.

۵۷ـ فی ابطال ظنون فاسده خطرت علی قلوب بعض الاطباء فی أمر الکواکب الحادثه فی الجو.
(در مورد باطل کردن خیالهای فاسد که بر خاطر برخی از پزشکان در مورد ستارگان پدید شونده در فضا خطور کرده است)

57. *Fi ibtal zunun fasida khatrat ala qulub bad al-atibba fi amr al-Kawakib al-haditha fi 'l-jaw.* (Nullification of the Whims of Some of the Physicians Arising in Their Mind Due to Appearance of the Planets in the Atmosphere).

This work is in 70 folios. Certain physicians interpreted as bad luck the appearance of the planets in the atmosphere. Biruni seeks to eliminate such false beliefs about stars.

۵۸_ مقاله فی الکلام علی کواکب ذوات الاذناب و الوائب.
(مقاله ای در مورد سخن در بارهٔ ستارگان دنباله دار و یالدار)

58. *Maqala fi 'l-kalam ala-kawakib dhawat al-adhnab wa'l dhawa'ib.* (Discourse on Planets with Tails and "Hair" (Comets and Lion's Mane).

This treatise deals with comets with "hair" and tails. This work was written in 65 folios.

۵۹_ مقاله فی مضیات الجو الحادثه فی العلو.
(مقاله ای در مورد پرتوهای پدیدار شوندهٔ فضا در آسمان)

59. *Maqala fi mudi'at al-*jaww al-haditha fi'l-*ulw.* (Discourse on Objects Illuminating the Atmosphere Which Appear in the Sky).

This treatise gives details of illuminating objects that appeared in the atmosphere. The number of folios is not mentioned in our sources.

۶۰_ مقاله فی تصفح کلام ابی سهل القوهی فی الکواکب المنقضه.
(مقاله ای در مورد جستار سخن ابوسهل کوهی در بارهٔ شتارگان شهابی)

60. *Maqala fi tasaffuh kalam Abi Sahl al-Quhi fi 'l kawakib al-munqadda.* (Examination of the Statements of Abu Sahl al-Quhi on the Falling Stars (Meteorites)).

This treatise is based on al-Quhi's information on falling stars and is an enlargement of his work. It is in 15 folios.

۶۱_ کتاب فی تحقیق منازل القمر.
(کتابی در مورد پژوهش منازل ماه)

61. *Kitab fi tahqiq manazil al-Qamar.* (Verification of the Mansions of the Moon).

This book is written in 180 folios. It deals with different locations of the moon in the sky.

۶۲_ فی الفحص عن نوادر ابی حفص عمر بن الفرخان.
(در مورد بازیابی نوادر ابوحفص عمر بن فرخان)

62. *Fi 'l-*fahs *an nawadir Abi Hafs Umar b. al-Farrukhan.* (Examination of the Curios in Possession of Abu Hafs Umar b. al-Farrukhan).

This work describes the research carried out by Biruni on the rare collections of

Abu Hafs Umar b. al-Farrukhan, one of very few Medieval scholars who worked towards discrediting astrology . This book had 240 folios.

٦٣ـ مقاله فی النسب التی بین الفلزات و الجواهر فی الحجم.
(مقاله ای در مورد نسبتهایی که میان فلزات و سنگهای گرانبها
از لحاظ حجم وجود دارد)

63. *Maqala fi l-nisab allati bayn al-filizzat wa 'l-jawahir fi 'l-hajm.* (Discourse on the Relationship Between the Specific Density of Metals and Precious Stones).

This work deals with specific density of some metals and precious stones. Biruni's findings are very similar to modern values. It was in 30 folios. Manuscript: Beirut 223 (L. Cheikho's catalogue, pp. 20-55).

٦٤ـ مقاله فی استخراج الاوتار فی الدائره بخواص الخط المنحنی فیها.
(مقاله ای در مورد محاسبهٔ وترها در دایره ها بوسیلهٔ خاصیتهای خط
منحنی که در آنهاست)

64. *Maqala fi istikhraj al-awtar fi 'l-al-da'ira bi-khwass al-khatt al-munhani fiha.* (Calculations of the Chords of the Circle by the Properties of its Curved Lines).

This treatise is about calculating the chord of a circle by the properties of its curved lines. It was written in 80 folios. Manuscript: Bankipore 2468/42=Patna 2,336,2519/40.—Cairo, 5,203.—Leiden, Or, 513 (5)=cc01012.—Murat Mulla, Istanbul, 1396/14.

٦٥ـ تذکره فی المساحت للمسافر المقوی.
(تذکره ای دربارهٔ زمین پیمایی برای مسافر قوی)

65. *Tadhkira fi l-misahat li 'l-musafir al-Muqawwi.* (Memoir on the Distance Covered by a Healthy Man).

This memoir was prepared to estimate how much distance a healthy man could cover on foot. This memoir is in 10 folios.

٦٦ـ مقاله فی نقل خواص الشکل القطاع الا ما یغنی عنه.
(مقاله ای در مورد انتقال خاصیتهای شکل قطاع به آنچه
بی نیاز از آن شکل میسازد)

66. *Maqal fi naql khawass al-shakl al-qatta ila ma yughni anhu.* (Transfer of the Properties of the Transversal Figure Which is Indispensable).

This treatise deals with the Transversal Theorem of Menelaus. This work contained 20 folios.

۶۷_ مقاله فی ان لوازم تجزی المقادیر لا الی نهایه قریبت من امر الخطین اللذین یقربان و لا یلتقیان فی الاستبعاد.
(مقاله ای در مورد نتایج حاصل از تقسیم بی پایان مقدارها درست نزدیک به امر همان دو خطی است که به هم نزدیکند، اما از لحاظ دور بودن بهم نمیرسند)

67. *Maqala fi anna lawazim tajzi' al-maqadir la ila nihaya qaribat min amral-Khattayn al-ladhayn yaqraban wa-la yaltaqiyan fi 'l-istibad.* (Necessary Consequences of the Division in Infinity of Two Lines in Proximity to Each, But Never Meeting at Any Distant Point).

This treatise explains how two lines that are close to each other, but converging, can be divided into values. It discusses the Hyperbole and their Asymptotes. This treatise was written in 10 folios.

۶۸_ مقاله فی صفت أسباب السخونه الموجوده فی العالم و اختلاف أصول السنه.
(مقاله ای در مورد توصیف سببهای گرمای موجود در عالم و اختلاف فصلهای سال)

68. *Maqala fi sifat asbab al-sukhuna al-mawjuda fi l-alam wa-ikhtilaf usul al-sanah.* (Description of the Causes of Heat Found in the World and the Differences in the Seasons of the Year).

This treatise describes the causes generating heat, the differences in the seasons and the cultivation of crops. It was written in 45 folios.

۶۹_ مقاله فی البحث عن الطریقت المتعارفه المذکوره فی کتاب الآثار العلویه.
(مقاله ای در مورد بررسی روش شناختهٔ یاد شده در کتاب پدیده های آسمانی)

69. *Maqala fil-bahth an al-tariqat 'l-mutaarrafat' l-madhkura fi kitab al-athar al-ulwiya.* (Examination of the Known Method as Mentioned in *Kitab al-Athar al-ulwiya*).

This is a treatise in 40 folios. It was quoted in number 56 above. It is probably related to the *Book of Meteorites* by Aristotle.

۷۰_ المسائل البلخیه فی المعنی المتعلقه بانکسار الصناعت.
مسائل بلخیان دربارهٔ موضوع مربوط به شکسته شدن فن)

70. *Al-Masa'il al-Balkhiya fi mana 'l-mutaallaqa bi. inkisar al-sinaat.* (The Balkhi Problems Relating to the Meaning Given to the Breaking of the Art).

This book was written on Abu Zayd Ali b. Sahl al-Balkhi: philosophy of the great al-Kindi (801-873 C.E.). Al-Kindi was one of the orginal members of the "House of Wisdom" in Baghdad. He is known as a philosopher and also wrote about astronomy, mathematics and and geography. It covered 70 folios.

۷۱_ الجوابات عن المسائل الوارده من منجمی الهند.
(پاسخ های مسائلی که از منجمان هند رسیده است)

71. *Al-jawabat an al-masa'il al-warida min munajjim 'l-Hind.* (Replies to Questions Raised by the Indian Astronomers).

This book was written on Biruni's relations with Kashmiri and Indian astronomers. The number of folios of this work is not in our sources.

۷۲_ الجوابات عن المسائل العشر الكشميريه.
(پاسخ های ده مسالهٔ کشمیریان)

72. *Al-jawabat an al-masa'il al-ashr al-Kashmiriya.* (Replies to the Ten Questions Raised by the Kashmiris).

This book is on general astronomy. Biruni attempts to answer questions from scholars from Kashmir. The number of folios of this work is not known.

۷۳_ کتاب التفهیم لاوائل صناعت التنجیم.
(آموزاندن مبادی فن ستاره شناسی)

73. *Kitab al-tafhim li-awa'il sina'at al-tanjim.* (The Book of Instructions in the Elements of the Art of Astronomy).

The title of this book can be translated as "Book of Understanding Basic Art of Astrology." The number of folios of this work is not known. Probably, Biruni himself translated it into Persian from Arabic. This was done for one of his few female students by the name of Lady Raihanah, who did not know Arabic. Her name is the origin of Biruni's nickname, Abu Raihan. Lady Raihanah remains a mysterious presence in the male-dominated world of scholarship. It has five different sections about Geometry, Arithmetics, Conditions of the Earth and the Sky, Astrolabes, and Principles of Astronomy. This book was translated into English by R. Ramsay Wright as the *Book of Instruction in the Elements of the Art of Astrology*. London: 1934. This translation, in facsimile format, is based on the British Museum's Arabic manuscript number 8349. Since Biruni did not believe in "Astrology" and the book contains nothing related to the modern sense of the word, we are convinced that it should be replaced by "Astronomy." Manuscript: Aligarh 121/11 (in Persian, 2 copies), No. 44, 45.—A. Zaki Basha

(Cairo).--Berlin No. 5666 (Arabic).—Bodleian, 1,221.—Bodleian, 2, 262 (Arabic).—British Museum Oriental 8349.—British Museum Persian 2, 451-452.—Feyz 1333.—Nuri Osmaniyeh (Istanbul) Persian 2780.—Paris 2497 (translated from Persian).—Paris Persian 49.—Rabat 439.—Topkapi Serai 3477/8.—Tehran 2,162.—Zaw S. Hamza hesp. 18/5.

۷٤ــ مقاله فی تقسیط القوه و الدلالات بین اجزاء البیوت الاثنا عشر.
(مقاله ای در مورد تقسیم همسان قوا و دلالتها میان اجزای خانه های دوازدهگانهٔ منطقه البروج)

74. *Maqala fi taqsit al-quwa wa'l-dalalat bayn ajza al-Buyut al-ithna ashar.* (Division of the Forces and Signs Between the Various Parts of the Twelve Mansions).

This work is an essay in 15 folios. It deals with the forces and signs that can be divided into twelve mansions of the Zodiac.

۷٥ــ مقاله فی حکایت طریق الهند فی استخراج العمر.
(مقاله ای در مورد بازگویی شیوهٔ هندیان در محاسبهٔ طول عمر)

75. *Maqala fi hikayat tariq al-Hind fi istikhraj al-umr.* (On the Account of the Indian Method of Calculating the Age).

This work is a treatise on a problem related to the equalization of the Indian method of calculating "the Age." See numbers 102 and 105 below. The number of folios of this work is not given in our sources.

۷٦ــ مقاله فی سیر سهمی السعاده و الغیب.
(مقاله ای در مورد سیر سهم سعاده و سهم غیب)

76. *Maqala fi sayr sahmay al-saadat wa'l-ghayb.* (Journey of the Two Arrows of Good Luck and Their Absence).

This work also relates to equalization, as does number 75 above. The number of folios of this work is not known.

۷۷ـ فی الارشاد علی تصحیح المبادی اشتمل علی النموذرات.
(در مورد رهنمونی به تصحیح اصولی که دربردارندهٔ زایچه هاست)

77. *Fi 'l-irshad Ala tashih al-mabadi ' ishtamala ala l-Namudharat.* (Guide to the Correction of the Principles Contained in the Namudhars).

The translation of the word "Namudhars" is "Horoscope." This work was written in 50 folios.

۷۸ـ مقاله فی تبین رأی بطلمیوس فی سالخداه.
(مقاله ای در مورد بیان کردن نظر بطلمیوس در بارهٔ سالخداه)

78. *Maqala fi tabyin r 'y Batlamiyus fi 'l-salkhudah.* (Elucidation of the Opinion of Ptolemy on Salkhudah).

This treatise is composed of only 7 folios. "Sal-khoda" is a Persian word that means "the King of the Year." In astrology, it was used as a reference to a lucky person with honor and dignity.

۷۹ـ ترجمه کتاب المواليد الصغیره لبراهیمهر.
(ترجمهٔ کتاب زایچه های کوچک از براهیمهر)

79. *Tarjamat Kitab al-Mawalid al-saghira li-Brahimihira.* (Translation of the Small Work of Varahamihira on Births).

This work describes Indian astronomy and astrology. The number of folios of this work is not known.

۸۰ـ ترجمه قصه وامق و عذرا.
(ترجمهٔ داستان وامق و عذرا)

80. *Tarjamat qissat Wamiq wa Adhra '.* (Translation of the Story of Wamiq and 'Adhra').

This is a translation of a love story that is believed to have been composed by a Persian poet by the name of Abul Qasim Unsuri (968-1040 C. E.), a contemporary of Biruni. Unsuri served as the poet laureat for the court of Sultan Mahmud and his son Sultan Masud. The number of folios is not known.

۸۱ـ ترجمه حدیث قسیم السرور و عین الحیات.
(ترجمهٔ داستان قسیم السرور و عین الحیات)

81. *Tarjamat hadith Qasim al-surur wa Ayn al-Hayat.* (Translation of the Love Story of Qasim al-Surur and Ayn al-Hayat*)*.

This is also a translation of a love story between Qasim and Ayn al-Hayat. In Arabic Ayn means eye and Hayat means life. The number of folios is not known.

۸۲ـ ترجمه حدیث اورمزدیار و مهریار.
(ترجمهٔ داستان اورمزدیار و مهریار)

82. *Tarjamat hadith Urmuzdyar wa Mihryar.* (Translation of the Story of Urmuzdyar and Mihryar).

This is a translation of the story of Urmuzd or Hormuz and Mihr. These two had divine powers and represent Jupiter and the Sun. The number of folios is not given in our sources.

۸۳ـ ترجمه حدیث صنمی البامیان.
(ترجمهٔ داستان دو بت بامیان بلخ)

83. *Tarjamat hadith Sanamay al-Bamiyan.* (Translation of the Story of the Two Idols at Bamiyan).

Bamiyan was the capital city of a great district with the same name that is located in what is now central Afghanistan. This book is about two large sculptures of Buddha cut in the live rock of the hillside, from the base to the summit. They were carved into the sandstone in the valley of Bamiyan in the Hindu Kush mountain ranges as early as the third century. At 53 and 36 meters, 175 and 120 feet, these statues were the tallest standing statues of Buddha in the world. They were considered to be the most remarkable and the most priceless representation of the Buddha on Earth. Unfortunately, in March of 2001, the Taleban Regime in Afghanistan caused an international outrage by destroying these statues. The number of folios is not known.

۸٤ـ ترجمه حدیث دادمه و کرامی دخت، بنت جهل الوادی.
(ترجمهٔ داستان دادمه و کرامی دخت، بنت جهل الوادی)

84. *Tarjamat hadith Dadhmeh wa-Kirami, dukht Jahli al-Wadi.* (Translation of the Story of Dadhmeh and Kirami, Daughter of Jahli al-Wadi).

Translation of another love story between Dadhmeh and Kirami. Dadhmeh is not identified but Kirami refers to the Persian word "Girami" meaning "Dear." The number of folios is not mentioned in our sources.

۸۵ـ ترجمه حدیث نیلوفر فی قصه دبیستی و بربهاکر.
(ترجمهٔ داستان نیلوفر در قصه دبیستی و بربهاکر)

85. *Tarjamat hadith Nilufar fi qissat Dabisti wa-Barbuhakar*. (Translation of the Anecdote of Nilufar in the Story of Dabisti and Barbuhakir).

Translation of a love story between Nilufar and Barbuhakar. Nilufar, the blue lotus, is commonly used as a female name among the Persian-speaking people. Barbuhakar may stand for "Prabhakar", a common male name in India. The number of folios is not mentioned in our sources.

۸٦ـ ترجمه قافیه الالف من الاتمام فی شعر ابی تمام.
(ترجمهٔ شعرهای دارای قافیهٔ الف از انجامین بخش ابو تمام)

86. *Tarjamat qafiyat al-alif min al-itmam fi shir Abu Tammam.* Translation of the Qafiyat al-Alif of the Last Section of Abu Tammam's Anthology of Verses).

This is a tranlation of Abu Tammam's anthology of verses. The last chapter of Abu Tammam's book is entitled "Babal-Hija." The last word of each verse ends with the letter "alif." The number of folios is not known.

۸۷ـ ترجمه مقاله فی الاستبحار فی قد الاشجار.
(ترجمهٔ مقاله در مورد تبحر یافتن در برش طولی درختان)

87. *Tarjamat maqala fi 'l-istibhar fi qad al-a ashjar.* (Translation of the Discourse on the Growth of the Sizes of the Trees).

This work is a translation of an essay on the growth of trees. The number of folios is not known.

۸۸ـ ترجمه تحصیل الراحه بتصحیح المساحه.
(ترجمهٔ حاصل کردن راحت با تصحیح مساحت)

88. *Tarjamat tahsil al-raha bi-tashih al-misaha.* (Translation of "Acquiring Rest by Correcting Land-Survey").

This is a translation of a plan for land measurement that can be done easily. The number of folios is not known.

۸۹ـ ترجمه التحذير من قبل الترک.
(ترجمهٔ دوری جستن از قبایل ترک)

89. *Tarjamat al-tahdhir min qibal al-Turk.* (Translation of "Guarding Oneself Against the Turks").

The aim of this translation was to remove the apprehensions that people had about the Turks. The number of folios is not known.

۹۰ـ ترجمه القرعه المصرحه بالعواقب.
(ترجمهٔ فال مشخص نشان دادن نتایج)

90. *Tarjamat al-qura al-musarrah bi 'l-awaqib.* (Translation of "The Lot Clearly Indicating the Consequences").

The number of folios is not known. Biruni classified this work under anecdote, translations of stories from Persian or Sanskrit.

۹۱ـ ترجمه القرعه المثمنه لاستنباط الضمائر المخمنه و شرح مزامير القرعه المثمنه.
(ترجمهٔ فال هشت سویی برای دریافتن نهانهای مظنون و شرح نای و نی فال هشت سویی)

91. *Tarjamat al-qura l-muthammana li-istinbat al-dama'ir al-Mukhammana wa-sharh mazamir al-qura l-muthammana.* (Translation of "The Octuple Lot" for Knowing the Hidden Intentions, and Commentary on "The Octuple Lot").

This is a translation of a book dealing with telling the future by using the "Octuple Lot." The number of folios is not known.

۹۲ـ ترجمه کلب یاره و هو مقاله للهند فی الامراض التی تجری مجرا العفونه.
(ترجمهٔ سگ یاره، مقاله ای است به زبان هندی، در بارهٔ بیماریهایی که عفونت ببار می آورند)

92. *Tarjamat Kalb-Yarah wa-huwa maqala li 'l-Hind. fi 'l-amrad allati tajri majra al-Ufuna.* (Translation of Kalb-Yara, which is an Indian Treatise on Diseases Which Spread Putrefaction).

This is a treatise on Indian septic or contagious diseases, their occurrences and their cure. The number of folios is not known.

۹۳ـ کتاب فی تحقیق ما للهند من مقاله مقبوله فی العقل مرذوله.
(کتاب پژوهش آنچه مربوط به هند، از پیرفته و نپیرفته در نظر عقل)

93. *Kitab fi tahqiq ma li'l-Hind min maqalatin maqbulatin fi'l-aql au mardhula.* (Verification of What Is Said About India Which Is Accepted or Rejected by Reason).

This is Biruni's well-known research work on India. It is the best account of India's religion, culture, science, philosophy and other aspects of Indian life in the Middle Ages. This is the world's first research compendium on Hindu civilization. Biruni gathered information for this book during Sultan Mahmud's military campaign in India. It was first translated by Edward Sachau into English in 1888 and was called *Al-Biruni's India: An Account of the Religion, Philosophy, Literature, Geography, Chronology, Astronomy, Customs, Laws and Astrology of India about A. D. 1030*. London: Trench Trubner. This book had 700 folios. The only doctoral dissertation on any work by Biruni was written by M. Ghayasuddin, *A Critical Analysis of the Writings of Abu'l Raihan Muhammad al-Biruni, Pertaining to India.* Unpublished Dissertation for Ph. D. degree in Arabic, Aligarh Muslim University, Aligarh, India, 1968.

۹٤ـ مقاله فی علت علامات البروج فی الزیجات من حروف الجمل.
(مقاله ای در مورد علت علامت گذاری های برجها در زیجها
بوسیلهٔ حروف ابجد)

94. *Maqala fi Illat alamat al-Buruj fi'l-zijat min huruf l-jummal.* (Reason of the Signs of the Zodiac in the Astronomical Tables by the Letters of Jummal).

This is a treatise in 15 folios. The signs of the Zodiac in the astronomical tables are represented by their numerical values.

۹۵ـ کلام فی المستقر و المستودع.
(سخنی در بارهٔ قرارگاه و امانتگاه)

95. *Kalam fi 'l-mustaqarr wa'l-mustawda.* (Discussion on al-Mustaqarr and al-Mustawda*)*.

This work is in 10 folios. In this title the word "Mustaqarr" means settled and fixed and the word "Mustawda" means consignment and deposit in Arabic. This book was listed under the category of customs and religions by Biruni himself.

۹۶ـ مقاله فی باسدیو الهند عند مجیئه الادنا.
(مقاله ای در مورد باسدیو هند هنگام نزدیک شدنش)

96. *Maqala fi Basdiw al-Hind ind maji 'ihi 'l-adna.* (Discourse on Basdiw of India and His Arrival in the Near Future).

This work deals with the Indian belief about the appearance of "Vasudeva" under ordinary conditions. The number of folios is not known.

۹۷ـ ترجمه کتاب شامل فی الموجودات المحسوسه و المعقوله.
(ترجمهٔ کتابی جامع در بارهٔ موجودات محسوس و معقول)

97. *Tarjamat Kitab Shamil fi maujudat al-mahsusa wa 'l-maqula.* (Translation of a Comprehensive Book on the Sensible and Reasonable Beings).

The number of folios is not known. This book was listed under the category of customs and religions by Biruni.

۹۸ـ ترجمه کتاب باتنجل فی الخلاص من الارتباک.
(ترجمهٔ کتاب پاتنجلی در رهایی از بند و گرفتاری)

98. *Tarjamat kitab Batanjali fi 'l-khalas min al-irtbak.* (Translation of the Book of Patanjali on Removing Confusion).

This is a translation of the book of Patanjali, a well-known Indian grammarian of the second century B. C. The number of folios for this book is not known. Manuscript: Kopruluzade 1589, fo. 408a/415a.

۹۹ـ التنبیه علی صناعت التمویه و هی احکام النجوم.
(هشدار دادن در بارهٔ فن مشتبه کردن که همان اختر گویی است)

99. *Al-tanbih ala sinaat al-tamwih wa-hiya ahkam al-nujum.* (Guarding Against the Art of Illusion Which Is Astrology).

This work deals with determination of the duration of human life as influenced by the planets. The number of folios is not known. Manuscript: Kopruluzade 1589.

۱۰۰ـ تنویر المنهاج الی تحلیل الازیاج.
(روشن کردن روشها به جهت تحلیل زیجها)

100. *Tanwir al-minhaj ila tahlil al-azyaj.* (Illumination of the Path to the Resolution of the Astronomical Tables).

This work functions as a sort of key to the astronomical tables. The number of folios is not mentioned in our sources.

۱۰۱ـ التطبیق الی تحقیق حرکت الشمس.
(تطبیق برای پژوهش حرکت خورشید)

101. *Al-Tatbiq ila tahqiq harkat al-Shams.* (Application for the Verification of the Monument of the Sun).

This work deals with the movements--especially the revolution--of the Sun. The number of folios is not mentioned in our sources.

۱۰۲ـ البرهان المنیر فی اعمال التسییر.
(دلیل روشن در مورد اعمال سیر دادن در مفهوم ستاره شناسی)

102. *Al-Burhan al-munir fi amal al-tasyir.* (Illuminating Proof Concerning the Operations of Equalization).

This book deals with chemical compounds. The number of folios is not mentioned in our sources.

۱۰۳ـ کتاب تنقیح التواریخ و امثال ذلک.
(ویراستن تاریخها و مانند آنها)

103. *Kitab tanqih al-tawarikh wa-amthal dhalika.* (Correction of Dates and Such Other Subjects).

This book deals with correction of dates and other astronomical problems. The number of folios is not mentioned in our sources.

۱۰٤ـ القانون المسعودی.
(قانون مسعودی)

104. *Al-Qanun al-Masudi.* (Canon Masudicus).

This work is Biruni's most complete astronomical research. It was dedicated to Sultan Masud, son of Mahmud of Ghaznavid Dynasty around the year 1030. The

Sultan was so pleased he offered Biruni an elephant load of silver coins. Biruni refused the reward and returned it to the treasury. Instead, he was satisfied with the facilities for research and other amenities that the Sultan provided for him for the rest of his life. The number of folios is not mentioned in our sources.

۱۰۵ـ الآثار الباقیه عن القرون الخالیه.
(آثار بجای مانده از سده های بگذشته)

105. Al-Athar al-Baqiya an al-Qurun al-Khaliya. (The Traces of the Past Centuries).

This was Biruni's first major research work, completed in the year 1000. It basically deals with chronology among ancient nations. A more precise meaning of the title is "Traces or Relics of the Leftover from the Empty Centuries." This book was translated from Arabic into English and edited with notes by Edward Sachau as the *Chronology of Ancient Nations*. London: W. H. Allen, 1879, xiv, 464 pp. It was also translated into Persian by A. Danasirisht and was published in Tehran in 1965. This book in Persian has 21 chapters and covers 611 pages. The number of original folios is not known.

۱۰۶ـ الارشاد الی ما یدرک ولا ینال من الأبعاد.
(رهنمونی به آنچه دریافته میشود اما به جهت دوری وصول بدان دست نمیدهد)

106. *Al-Irshad ila ma yudraku wa la yunalu min al-abad.* (Guide to the Distances Which Can Be Perceived But Can Not Be Reached).

This work describes the measurement of distances that are perceived, yet can not be covered. The number of folios is not known.

۱۰۷ـ الکتاب فی المکاییل و الموازین و شرایط الطیار و الشواهین.
(کتابی در بارۀ پیمانه ها و ترازوها و شرط های زبانۀ ترازو و شاهین ها)

107. *Al-Kitaba fi 'l-makayil wa 'l-mawazin wa-shara'it al-tiyar wa-l-shawahin.* (Writing on Measures and Balances and the Conditions of Tiyar and the Needles of the Balances).

This work deals with weights and measures and discusses accuracy of scales and other tools for measuring volume. The number of folios not known.

۱۰۸ــ جمع الطرق السائره فی معرفت أوتار الدائره.
(گردآوری راه و روشهای رایج در باب شناخت وترهای دایره)

108. *Jama al-turuq al-sa'ira fi marifat awtar al-da'ira.* (Collection of the Methods to Know the Chords of a Circle).

This work deals with the methods by which the chords of a circle can be calculated. The number of folios is not known.

۱۰۹ــ تصور أمر الفجر و الشفق فی جهتی الشرق و الغرب من الافق.
(تصور چگونگی سپیدی پگاه و سرخی شامگاه در دو جهت شرق و غرب از افق)

109. *Tasawwur amr al-fajr wa'l-shafaq fi jihatay al-sharq wa'l-gharb min al-ufuq.* (Representation of the Phenomena of Dawn and Evening Twilight in the Two Directions of East and West on the Horizon).

This work deals with the rising and setting of the Sun in the east and west. The number of folios is not known.

۱۱۰ــ تکمیل صناعه التسطیح.
(کامل کردن فن نقشه کشی)

110. *Takmil Sinaat al-tastih.* (Perfection of the Art of Plane Projection).

This work describes thoroughly the Glider Plane Projection. The number of folios is not known.

۱۱۱ــ جلاء الأذهان فی زیج البتانی.
(روشن کردن اذهان در مورد زیج بتانی)

111. *Jala' al-adhan fi Zij al-Battani.* (Clearing of the Mind with Regard to the Astronomical Tables of al-Battani).

This work deals with the Zij, Astronomical Tables, of the famous astronomer al-Battani (Died 929 C. E.). His Zij was more accurate than all others calculated up to that time, but was still scrutinized by Biruni. The number of folios is not known.

۱۱۲ــ تحدید المعموره و تصحیحها فی الصوره.
(تعیین حدود آبادان، زمین مسکون، و تصحیح آن در نقشه)

112. *Tahdid al-mamura wa tashihuha fi'l-sura.* (Delimitation of the Inhabited World and Its Correction on the Map).

This work deals with the delimitation and determination of the boundaries of countries and cities as well as with the correction of their positions on the map. The number of folios is not known.

۱۱۳ـ علل زیج جعفر المکنی بابی معشر.
(شرح برهانی زیج جعفر مشهور به ابومعشر بلخی)

113. *Illal zij jafar al-Mukanna bi-Abu Mashar.* (Defects of the Astronomical Table of Jafar, Surnamed Abu Mashar).

In this book Biruni strongly disputes the calculations of the cycles made by the celebrated astronomer Abu Mashar (805-885 C. E.). The number of folios is not known.

۱۱٤ـ کتاب فی السموت.
(کتابی در مورد سمتها)

114. *Kitab fi 'l-sumut.* (Book on Azimuths).

This book was composed in Biruni's name by Abu Nasr Mansur Ibn Ali Ibn Iraqq. Abu Nasr was a member of the royal family of the kingdom of Khwarazm. He was Biruni's teacher, colleague, and friend. Biruni calls him his "Master." The number of folios is not known.

۱۱۵ـ کتاب فی علت تنصیف التعدیل عند اصحاب السندهند.
(کتابی در مورد علت نیم کردن تعدیل در نزد سندهند گرایان)

115. *Kitab fi illat tansif al-tadil ind ashab al-Sindhind.* (Reasons for the Division in Two of the Equations According to the Authors of Sindhind *(Siddhanta)).*

Same as number 114 above. The number of folios is not known.

۱۱٦ـ کتاب فی تصحیح کتاب ابراهیم بن سنان فی تصحیح اختلاف الکواکب العلویه.
(کتابی در مورد تصحیح کتاب ابراهیم بن سنان در تصحیح ناهمانندی اختران آسمانی)

116. *Kitab fi tashih Kitab Ibrahim b. Sinan fi tashih ikhtilaf al-kawakib al-Ulwiyya.* (On the Correction of the Book of Ibrahim b. Sinan on the Rectification of the Differences in the Celestial Planets).

Same as number 114 above. ~~The number of folios is not known.~~

۱۱۷_ رساله فی براهین اعمال حبش بجدول التقویم.
(رساله ای در مورد دلایل کارهای حبش در جدول تقویم)

117. *Risala fi barahin amal Habash bi-jadwal al-taqwim.* (Proofs of the Work of Habash on the Tables of Rectification (of the Calendar)).

Same as number 114 above. This tract was published along with 14 others of Abu Nasr Mansur, in a single volume entitled *Rasa'il Abu Nasr ila'l-Biruni (*see numbers 118-125, 178, and 179 below). Habash's nickname was "Hasib" which in Arabic means "One Who Can Calculate." He calculated tables of sines, tangents, and standard astronomical functions. Habash was born in the early 9th century in Turkmenistan but worked in Baghdad and died in 870 C. E. The number of folios is not known.

۱۱۸_ رساله فی تصحیح ما وقع لأبی جعفر الخازن من السهو فی زیج الصفائح.
(رساله ای در مورد تصحیح اشتباهاتی که ابوجعفر خازن را در زیج الصفائح رخ داده است)

118. *Risala fi tashih ma waqaa li Abu Jafar al-Khazin min al-sahw fi zij al-safa'ih.* (Corrections of the Omissions of Abu Jafar al-Khazin in the Disk Tables).

Same as number 114 above. This tract was published along with 14 others of Abu Nasr Mansur, in a single volume entitled *Rasa'il Abu Nasr ila'l-Biruni (*see numbers 118-125, 178, and 179 below). Abu Jafar al-Khazin (900-971 C. E.) was a geographer-astronomer. He wrote a commentary on Ptolemy's *Almagest* which was criticised by Biruni for being too verbose and repetitve. The number of folios is not known. Manuscript: Bankipore 2468/8.

۱۱۹_ رساله فی مجازات دوائر السموت فی الاصطرلاب.
(رساله ای در مورد گذارهای دایره های سمتها در اسطرلاب)

119. *Risala fi majazat dawa'ir al-sumut fi l-asturlab.* (The Passage of the Azimuth Circles in the Astrolabe).

Same as number 114 above. This tract was published along with 14 others of Abu Nasr Mansur, in a single volume entitled *Rasa'il Abu Nasr ila'l-Biruni (*see numbers 118-125, 178, and 179 below). The number of folios is not known.

١٢٠ـ رساله فی جدول الدقائق.
(رساله ای در مورد جدول دقیقه ها)

120. *Risala fi jadwal al-daqa'iq*. (The Table of the Minutes).

Same as number 114 above. This tract was published along with 14 others of Abu Nasr Mansur, in a single volume entitled *Rasa'il Abu Nasr ila'l-Biruni (*see numbers 118-125, 178, and 179 below). The number of folios is not known.

١٢١ـ رساله فی البراهین علی عمل محمد بن الصباح فی امتحان الشمس.
(رساله ای در مورد دلایل بر کار محمد بن صباح در بارهٔ آزمودن خورشید)

121. *Risala fi'l-burahin ala amal Muhammad bin al-Sabbah fi imtihan al-Shams.* (The Proof, Concerning the Work of Muhammad b. al-Sabbah on the Examination of the Sun).

Same as number 114 above. This tract is published along with 14 others of Abu Nasr Mansur, in a single volume entitled *Rasa'il Abu Nasr ila'l-Biruni (*see numbers 118-125, 178, and 179 below). This work by Muhammad bin al-Sabbah, appears to be his writing on the method for determining the mid-day. The number of folios is not known.

١٢٢ـ رساله فی الدوائر التی تحد الساعات الزمانیه.
(رساله ای در مورد دایره هایی که حد ساعات زمانی را معین میکنند)

122. *Risala fi'l-dawa'ir allati tahuddu al-saat al-zamaniyya.* (On the Circles Which Limit the Temporal Hours).

This tract was published along with 14 others of Abu Nasr Mansur, in a single volume entitled *Rasa'il Abu Nasr ila'l-Biruni (*see numbers 118-125, 178, and 179 below). The number of folios is not known.

١٢٣ـ رساله فی البرهان علی عمل حبش فی مطالع السمت فی زیجه.
(رساله ای در مورد دلیل بر کار "حبش حاسب" در بارهٔ پدیدگاههای سمت در زیجش)

123. *Risala fi'l-burhan ala amal Habash fi matali al-simt fi Zijihi.* (The Proof Concerning the Work of Habash on the Ascensions of the Azimuth in His Tables).

This tract was published along with 14 others of Abu Nasr Mansur, in a single volume entitled *Rasa'il Abu Nasr ila'l-Biruni (*see numbers 118-125, 178, and

179 below). For some basic information about Habash, please see entries 4 and 117. The number of folios is not mentioned in our sources.

۱۲٤ـ رساله فی معرفت القسی الفلکیه بطریق غیر طریق النسبت المولفه.
(رساله ای در مورد شناخت قوسهای فلکی با روشی بجز روش نسبت مولفه)

124. *Risala fi marifat al-qusiyy al-falakiya bi-tariq ghayr tariq al-nisbat al-muallafa.* (The Knowledge of the Spheric Arc by a Method other Than the One Which Is Composed).

This tract was published along with 14 others of Abu Nasr Mansur, in a single volume entitled *Rasa'il Abu Nasr ila'l-Biruni (*see numbers 118-125, 178, and 179 below). The number of folios is not mentioned in our sources.

۱۲۵ـ رساله فی حل شبهت عرضت فی الثالثة عشر من کتاب الاصول.
(رساله ای در مورد حل یک اشکال که در سیزدهمین گفتار از کتاب اصول اقلیدس رخ داده است)

125. *Risala fi hall shubhatin aradat fi'l-thalithata ashrata min kitab al-'Usul.* (Solution of a Problem Appearing in the Thirteenth Tract of the "Book of Principles").

This tract was published along with 14 others of Abu Nasr Mansur, in a single volume entitled *Rasa'il Abu Nasr ila'l-Biruni (*see numbers 118-125, 178, and 179 below). This work deals with difficult problems of geometry. The number of folios is not known. Manuscript: Bankipore 2468/18.

۱۲٦ـ کتاب فی مبادی الهندسه.
(کتابی در مورد مبادی هندسه)

126. *Kitab fi mabadi' al-handasa.* (The Principles of Geometry).

This work was composed in Biruni's name by Abu Sahl Isa b.Yahya al-Masihi. The number of folios is not known.

۱۲۷ـ کتاب فی رسوم الحرکات فی الاشیاء ذوات الوضع.
(کتابی در مورد قوانین حرکت در اشیاء دارای وضعیت)

127. *Kitab fi Rusum al-harakat fi'l-'ashya' dhawat al-wad.* (Rules of the Movements of Things with Positions).

The same as number 126 above. The object of the tract is not clear from the title. The number of folios is not known.

۱۲۸ــ کتاب فی سکون الارض اوحرکاتها.
(کتابی در مورد سکون یا حرکت زمین)

128. *Kitab fi sukun al-ard au harakatiha.* (On the Fixed Position of the Earth or Its Movement).

The same as number126 above. Biruni frequently deals with the question of the movement of the Earth. The number of folios is not known.

۱۲۹ــ کتاب فی التوسط بین ارسطوطالیس و جالینوس فی المحرک الاول.
(کتابی در مورد میانجیگری ارسطو و جالینوس در مورد محرک نخستین)

129. *Kitab fi 'l-tawassut bayn Aristutalis wa Jalinus fi 'l-muharrik al-awwal.* (Mediation Between Aristotle and Galen on the Subject of the "Prime Mover").

The same as number126 above. This work deals with the views of Aristotle and Galen on the first cause of the movement of the Universe. The number of folios is not known.

۱۳۰ــ رساله فی دلالت اللفظ علی المعنی.
(رساله ای در مورد دلالت لفظ بر معنی)

130. *Risala fi dalalat al-lafz ala'l-ma'na.* (Epistle on the Word Indicating the Sense).

The same as number126 above. This work deals with logic and discusses the importance of the word in its fullest sense. The number of folios is not known.

۱۳۱ــ رساله فی سبب برد ایام العجوز.
(رساله ای در مورد سبب سرمای ایام عجوز، هفت روز پایان زمستان)

131. *Risala fi sabab bard ayyam al-ajuz.* (Cause of Cold During the Days of the Old Woman).

The same as number126 above. This book deals with the recurrence of cold for a week at the end winter in the month of March. The number of folios is not known.

۱۳۲ـ رساله فی علت التربیه التی تستعمل فی احکام النجوم.
(رساله ای در مورد انگیزۀ تربیت که در احکام نجوم بکار میرود)

132. *Risala fi illat al-tarbiya (?) allati tustamalu fi ahkam al-Nujum.* (The Reason of Improvement (?) Utilized in Astronomy).

The same as number126 above. This work is about one's progress and improvement in learning (?) astronomy. The number of folios is not known.

۱۳۳ـ رساله فی آداب صحبت الملوک.
(رساله ای در مورد آیین های همنشینی با شاهان)

133. *Risala fi Adab suhbat al-muluk.* (On Good Manners in the Company of Kings).

The same as number126 above. This book deals with manners and etiquette to be observed when one is in the company of a king. The number of folios is not known.

۱۳٤ـ رساله فی قوانین الصناعت.
(رساله ای در مورد قانونهای فن و هنر)

134. *Risala fi qawanin al-*sina'a. (On Canons of the Art).

The same as number126 above. This work is about the rules of astronomy. The number of folios is not given in our sources.

۱۳۵ـ رساله فی دستور الخط.
(رساله ای در مورد راهنمایی خط و خوشنویسی)

135. *Risala fi dastur al-khatt.* (Rules of Calligraphy).

The same as number126 above. This was probably dedicated to Sultan Mawdud (1040-1048), the grandson of Sultan Mahmud of the Gaznavid Dynasty. It is about the art of calligraphy. The number of folios is not given in our sources.

۱۳٦ـ رساله فی الغزلیات الشمسیه.
(رساله ای در مورد طلوعات خورشیدی)

136. *Risala fi 'l-ghazaliyat al-shamsiya.* (The Rising of the Sun).

The same as number126 above. The word *Ghazaliyat* is not translated as lyrics or odes, but read as *Ghazalat*, meaning the "Rising Sun." It has been classified as a book in geodesy. The number of folios is not given in our sources.

١٢٧ـ الرساله النرجسیه.
(رسالهٔ نرگس نامه)

137. *Al-Risala al-narjisiya.* (The Narcissus Epistle).

The same as number126 above. This work is listed under general astronomy. The number of folios is not mentioned in our sources.

١٣٨ـ الرساله المعنونه بمن و عن.
(رساله ای به عنوانهای من و عن)

138. *Al-Risala al-ma'nuna bi-min wa-an.* (The Epistle Entitled "min wa an").

This book was written in Biruni's name by Abu Ali Hassan Ibn Ali Jili. This treatise is on grammar, specifically on the comparative significance of the Arabic prepositions "min" and "an." The number of folios is not known.

١٣٩ـ استیعاب فی تسطیح الکره.
(یک بررسی کامل در بارهٔ تصویر کردن کره)

139. *Istiab fi tastih al-kura.* (Exhaustive Study on the Plane Projection of the Sphere).

This work is listed under stereography and deals with the plane projection of spheres. The number of folios is not known.

١٤٠ـ الزیج العلائی.
(زیج علائی)

140. *Al-Zij al-Ala'i.* (The Astronomical Tables of al-Ala'i).

This work is probably written by Biruni. However, some scholars believe that it was composed by either Muayyadin Urdi or, even more likely, by Aladin Nishaburi. The number of folios is not known.

١٤١ـ الزیج المسعودی.
(زیج مسعودی)

141. *Al-Zij al-Masudi.* (The Masudian Astronomical Tables).

Perhaps the same as *Al-Qanun al-Masudi*, Canon Masudicus (see number 104 above). The number of folios is not known.

١٤٢_ كتاب تسطيح الكره.
(كتاب مسطح كردن كره)

142. *Kitab tastih al-kura.* (Plane Projection of the Sphere).

This work is listed under general astronomy, similar to number 139, dealing with projection of spheres. The number of folios is not known.

١٤٢_ كتاب الدرر في سطح الكره.
(كتابي در مورد مرواريدها در بارهٔ رويهٔ كره ها)

143. *Kitab al-durar fi sath al-kura.* (The Pearls on the Surface of the Sphere).

Also listed under general astronomy, this work deals with spheres. The number of folios is not known.

١٤٤_ اختصار المجسطي.
(خلاصهٔ كتاب مجسطي، نوشتهٔ بطلميوس)

144. *Ikhtisar al-majisti.* (The Abridgment of *Almagest*).

This is Biruni's abridgement of Ptolemy's book *Almagest*. The number of folios is not known.

١٤٥_ كتاب في لوازم الحركتين.
(كتابي در مورد لوازم دو حركت)

145. *Kitab fi lawazim al-harakatayn.* (The Concomitants Necessary for the Two Movements).

This work deals with two movements of stars, one from the east to the west and the other from the west to the east. The number of folios is not known.

۱٤٦ــ کتاب الاستشهاد باختلاف الارصاد.
(کتابی در مورد گواه یابی به سبب اختلاف رصدها)

146. *Kitab al'istishhad bi'-ikhtilaf al-arsad.* (Testimony on Disagreements with Regard to Astronomical Observations).

This work is classified under general astronomy with some astronomical observations. The number of folios is not known.

۱٤۷ــ مسائل سأل عنها ابن سینا.
(مسائلی که بیرونی از ابن سینا پرسیده است)

147. *Masa'il sa'ala anha Ibn Sina.* (Questions Posed to Avicenna).

This book deals with the correspondence Biruni had with Avicenna, Ibn Sina (980-1037). The number of folios is not known.

۱٤۸ــ کتاب الشموس الشافیه للنفوس.
(کتاب خورشیدهای بهبود بخش جانها)

148. *Kitab al-shumus al-shafiya li'l-mufus.* (The Book of the Suns, the Healers of the Souls).

This work is listed under astrology. The number of folios is not known.

۱٤۹ــ ارشاد فی احکام النجوم.
(رهنمونی دربارهٔ احکام نجوم)

149. *Irshad fi ahkam al-nujum.* (Guide to Astrology).

This book is also listed under astrology. The number of folios is not known.

۱۵۰ــ کتاب تاریخ ایام السلطان محمود و اخبار ابیه.
(کتاب تاریخ روزگار سلطان محمود و اخبار پدرش)

150. *Kitab ta'rikh ayyam al-sultan Muhmud wa-akhbar abihi.* (History of the Times of Sultan Mahmud and the Annals of His Father).

This work is on the history of the infamous Sultan Mahmud (998-1030) of the Ghaznavid Dynasty and his father. The number of folios is not known.

۱۵۱_ فی صفیحه التسییر.
(در بارهٔ صفحهٔ تسییر)

151. *(Fi Safihat at-tasyir)*. (On Plate of Tasyir).

This work deals with the point of impact of light rays. The number of folios is not known.

۱۵۲_ کتاب المدخل الی علم النجوم و علی حدود النحوس و المواضع القاطعه بالتجارب.
(کتاب درآمد به علم نجوم و بر حدود بداختریها و جاهای قطع کننده با آزمایش)

152. *Kitab al-madkhal ila ilm al-nujum wa-ala hudud al-nuhus wa'l-mawadi' al-qatia bi'l-tajarib.* (Introduction to Astrology and On the Limits of the Unlucky Moments and the Decisive Places in the Light of the Experiences).

This work deals with the influence of stars on human activities based on Biruni's experience. The number of folios is not known.

۱۵۳_ کتاب العجایب الطبیعت و الغرائب الصناعت.
(کتاب شگفتیهای طبیعی و غرایب صنعتی و فنی)

153. *Kitab al-aja'ib al-tabiiya wa'l-ghara'ib al-sinaiya.* (The Marvels of Nature and the Wonders of the Arts).

This book was written about enchantments, charms, and talismans. The number of folios is not known.

۱۵٤_ کتاب تقاسیم الاقالیم.
(کتاب تقسیمات کشورها)

154. *Kitab taqasim al-aqalim*. (The Divisions of the Climes).

This work is listed under climatology. It is about classification of the known world into several types of climates. The number of folios is not known.

۱۵۵_ کتاب فی اعتبار مقدار اللیل و النهار بطریق تبعد عن موضعات المنجمین و ألقابهم.
(کتابی در مورد تامل در طول مدت شب و روز با روشی به دور از اصطلاحات و واژه های قراردادی منجمان)

155. *Kitab fi itibar miqdar al-layl wa 'l-nahar bi-tariq tabudu an muwadaat al-munajjimin wa-alqabihim.* (Determination of the Duration of Night and Day by a Method Which Is Far Removed from the Stands and Expressions of the Astronomers).

This book is classified under general astronomy and deals with determination of the duration of days and nights. The number of folios is not known.

١٥٦ــ الجماهر فی الجواهر.
(مجموعه ها در شناخت گوهرها)

156. *Al-Jamahir fi l-jawahir.* (The Collection on Gems and Precious Stones).

This work is listed under mineralogy. It deals with the properties of gems and precious stones and metals. Biruni described one hundred known minerals from around the world. He measured "specific Density" of some mineral and metals that are very similar to their modern values. This book was dedicated to Sultan Mawdud (1040-1048) of the Ghaznavid Dynasty. This book is believed to be the most complete source of information on precious stones written in the Middle Ages. The number of folios is not cited in our sources.

١٥٧ــ کتاب نزهت النفوس و الافکار فی خواص المواليد الثلاثه المعادن و النبات و الحجار.
(کتاب خرمی جانها و اندیشه ها در مورد ویژگیهای فرآورده های سه گانه کانی و گیاهان و سنگها)

157. *Kitab Nuzhat al-Nufus wa 'l-afkar fi khawas al-mawalid al-thalathat l-maadin wa 'l-nabat wa 'l-ahjar.* (Recreation of the Mind and the Thoughts on the Properties of the Three Species of Matter, Metals, Plants and Stones.)

This work also deals with mineralogy but explains properties of matter, metals, plants, and stones. The number of folios is not cited in our sources.

١٥٨ــ کتاب الصيدله فی الطب.
(کتاب دارو شناسی در پزشکی)

158. *Kitab al-Saydala fi 'l-tibb.* (The Book on Medicinal Drugs).

This is a comprehensive book on theoretical botany and pharmacy. Biruni deals with medical issues such as occurrence, medical usage, diagnosis, and natural drugs. Included were etymology, descriptions which were based on his experience and revisions of some Greek opinions. He provided the exact names of these drugs in Arabic, Greek, Hebrew, Persian, and Sanskrit. The book was completed after Biruni's death with the help of a physician friend who also had a minor role in gathering information and doing research with Biruni. The number of folios is not known.

١٥٩ـ كتاب الحجار.
(كتاب كانشناسی)

159. *Kitab al-Hijar*. (Mineralogy).

This title is classified under mining and mineralogy. It seems to be a translation of Aristotle's book on mineralogy into Arabic. The number of folios is not known.

١٦٠ـ كتاب المسامره فی اخبار خوارزم.
(كتاب داستانسرایی در بارهٔ اخبار خوارزم)

160. *Kitab al-Musamara fi Akhbar Khwarizm*. (Conversation on the Annals of Khwarizm).

This book was written in Persian by Biruni and contained a complete history of his homeland, Khwarazm, up to his time. The original version of this book is lost, but it is reproduced in its entirety by Bayhaqi in his book called *Tarikh Bayhaqi*, History of Bayhaqi. The number of folios is not known.

١٦١ـ ترجمة اخبار المقنع.
(ترجمهٔ اخبار هاشم بن حكيم معروف به المقنع از فارسی به عربی)

161. *Tarjamat Akhbar al-Muqanna*. (Translation of the Annals of Maqanna).

This is a translation from Persian into Arabic by Biruni himself. This book is about the history of a movement against Baghdad by Al-Muqanna. He was originally called Hashim Ibn Hakim Al-Muqanna, literally, the "Veiled One." He was also known as the "Masked Man of Khorasan" (between 777-780). He was a religious leader in the city of Merv in Khorasan. He claimed to be an incarnation of God and gave rise to a new heresy whose followers dressed in white and are known in history as "Sapid Jamagan." The number of folios is not known.

۱٦۲ــ كتاب فى اخبار المبيضه و القرامطه.
(كتابى در مورد اخبار سپيد جامگان و قرمطيان)

162. *Kitab fi Akhbar al-Mubayyida wa 'l-Qaramita.* (History of the Mubayyida and the Qaramita).

This work is a history of "White Dressed" people that followed Al-Muqanna (see number 161 above), and Qaramita or Qaramatians/Karmatians. The latter is the name of a sub-sect of the Shiah Islam, the Ismailiyah. The number of folios is not known.

۱٦۳ــ كتاب شرح شعر ابى تمام.
(كتاب شرح شعر ابو تمام)

163. *Kitab Sharh Shir Abi Tammam.* (Commentary on the Poetry of Abu Tammam.)

This work is listed under poetry and comments on the poetry of Abu Tammam. The number of folios is not known.

۱٦٤ــ كتاب التعلل بحالت الوهم فى معانى نظم اولى الفضل.
(كتاب توجيه انتقال صورت خيال در معنيهاى نظم فضلمندان)

164. *Kitab al-taallul bihalat al-wahm fi maani nazm ula 'l-fadl.* (Justification for the Transfer of the Imagination to the Significance of the Best Poetry).

Also listed under poetry, this book deals with the best available in Biruni's time. The number of folios is not known.

۱٦٥ــ كتاب مختار الاشعار و الآثار.
(كتاب گزيدهٔ اشعار و آثار)

165. *Kitab Mukhtar al-Ashar wa 'l-Athar.* (Selected Verse and Traditions).

This is another book classified under poetry and includes some selected verses from Biruni's time. The number of folios is not known.

۱٦٦ــ مجموعه من الاشعار.
(مجموعهٔ اشعار)

166. *Majmua min al-ashaar.* (Collection of Verses).

This work probably included some of Biruni's own poetry. The number of folios is not known.

۱۶۷ـ کتاب المعنون بالدستور.
(کتابی بنام عنوانها در دستور)

167. *Kitab al-Maamun bi'l-dastur*. (The Book Entitled *al-Dastur*).

This work is an exhaustive treatise on ethics. In Arabic, the word "Dastur" means direction and instruction. This book was dedicated to Sultan Mawdud (1040-1048) of the Ghaznavid Dynasty. The number of folios is not known.

۱۶۸ـ رساله فی فهرست کتب محمد بن زکریا الرازی.
(رساله ای در مورد فهرست کتابهای محمد بن زکریای رازی)

168. *Risala fi Fihrist kutub Muhammad b. Zakariyai al-Razi*. (On the List of the Books of Muhammad Ibn Zakariyai al-Razi).

As mentioned in the beginning of this section, this book is usually referred to as the *Fihrist*. From this book one finds the names of all of the books written by Razi but also Biruni provides the reader with 138 names of his own books. It is amazing to know that Biruni is the first and probably the only scholar in history who spent many years collecting and documenting another scholar's writings.

۱۶۹ـ حکایت الألت المسمات السدس الفخری.
(داستان ابزار موسوم به شش یک فخری)

169. *Hikayat al-alat'l-musammat al-sudus al-Fakhri (al-Fakhira)*. (Account of the Instrument Known as Sextant).

In this book, Biruni discusses a primitive form of sextant, an instrument for measuring angular distances, latitudes and longitudes, that was invented by Fakhrdin Khujandi. Biruni met with Khujandi in the city of Rayy before the end of the first millenium. Khujandi died in 1000. Khujandi's sextant was a very large and heavy instrument built on a mountain above the city of Rayy to observe longitude transits of the sun near the solstices. This book deals with Khujandi's astrolabe/sextant. In Rayy, Biruni, a poor young scholar, was not well received by the scholars because of his appearance. The number of folios is not known.

۱۷۰ـ ریاضت الفکر و العقل.
(ورزش اندیشه ها و خرد)

170. *Riyadat al-fikr wa'l-aql*. (Exercise of the Thought and the Reason).

This work deals with a method of self-discipline for improving thought and reasoning. The number of folios is not known.

۱۷۱ـ اجوبت المسائل الخوارزميه.
(پاسخهای مسائل خوارزمیان)

171. *Ajwibat al-masa'il al-Khwarizmiyya.* (Replies to the Khwarizmian Questions).

This book deals with Biruni's replies to Ibn Sina in Khwarazm. The number of folios is not mentioned in our sources. Manuscript: Asafiyyah 1.

۱۷۲. فی الاعتراض علی کتاب ابن سینا فی حجت الحق.
(اعتراض به ادعاییهٔ ابن سینا در حجت بودن قولش در بارهٔ حقیقت)

172. *Fi'l-I'tirad ala kitab Ibn Sina fi Hujjat al-Haqq.* (Objections to the Work of Avicenna on the Proof of the Truth).

This book seems to be a refutation and counter argument addressed to Avicenna, Ibn Sina. The number of folios is not mentioned in our sources.

۱۷۳ـ مجموعه من الحکم.
(مجموعه ای از حکمت ها)

173. *Majmua min al-hikam.* (Collection of Wise Sayings).

This book contains the wise sayings of Biruni as a great scholar. The number of folios is not mentioned in our sources.

۱۷٤ـ ترجمه کتاب سامخیه.
(ترجمهٔ کتاب سامخیه)

174. *Tarjamat Kitab Samkhiya.* (Translation of the Book on Sankhya).

This is a translation of Sankhya from Sanskrit into Arabic by Biruni. It seems to be related to number 97 above. The number of folios is not mentioned in our sources.

۱۷۵ـ ترجمه کتاب فی اصول الهندسه لاقلیدس الی لغت الهند.
(ترجمهٔ کتابی در اصول هندسهٔ اقلیدس به زبان هندی)

175. *Tarjamat Kitab fi Usul al-handasa li-Uqlidous ila lughat al-Hind.* (Translation in Sanskrit of the *Book of Elements* by Euclid).

This work is a translation of the *Book of Elements* of Euclid from Arabic into Sanskrit under Biruni's supervision. The number of folios is not known.

۱۷۶_ ترجمه کتاب المجسطی لبطلمیوس الی لغت الهند
(ترجمهٔ کتاب مجسطی بطلمیوس به زبان هندی)

176. *(Tarjamat Kitab al-Majasti li-Batlamiyus ila lughat al-Hind*). (Translation in Sanskrit of Ptolemy's *Almagest*).

This is a translation of Ptolemy's book *Almagest* from Arabic into Sanskrit by Biruni. The number of folios is not known.

۱۷۷_ ترجمه کتاب فی الاصطرلاب لابی ریحان الی لغت الهند.
(ترجمهٔ کتابی در بارهٔ اسطرلاب از ابوریحان به زبان هندی)

177. *Tarjamat Kitab fi 'l-Usturlab li-Abi Raihan ila lughat al-Hind*. (Translation in Sanskrit of Abu Raihan's book on Astrolabe).

This is a translation of Biruni's book on astrolabes from Arabic into Sankrit. The number of folios is not known.

۱۷۸_ رساله ابی نصر فی جواب مسائل الهندسه.
(رسالهٔ ابونصر در پاسخ مسائل هندسه)

178. *Risalat Abi Nasr fi Jawab masa'il al-handasa*. (The Epistle of Abu Nasr in Reply to the Problems of Geometry).

This tract was published along with 14 others of Abu Nasr Mansur, in a single volume entitled *Rasa'il Abu Nasr ila'l -Biruni (*see numbers 118-125, 178, and 179 below). The number of folios is not given in our sources.

۱۷۹_ فصل من کتاب لابی نصر فی کریت السماء.
(بخشی از کتاب ابونصر در بارهٔ کروی بودن آسمان)

179. *Fasl min kitab li-Abi Nasr fi Kurriyat al-Sama'*. (Chapter of a book of Abu Nasr on Sphericity of the Heavens).

This tract was published along with 14 others of Abu Nasr Mansur, in a single volume entitled *Rasa'il Abu Nasr ila'l -Biruni (*see numbers 118-125, 178, and 179 below). The number of folios is not mentioned in our sources.

۱۸۰ـ كتاب ابى الريحان الى ابى سعيد.
(نامهٔ ابوريحان به ابوسعيد)

180. *Kitab Abi'l-Raihan ila Abi Said.* (Writing of Abu Raihan to Abu Said).

This work deals with two theorems of spherical trigonometry of Abu Nasr Mansur. It seems to be an epistle sent to or dedicated by Abu Raihan Biruni to Abu Said al Sijzi (945-1020 C. E.). Little is known about Abu Said but fairly accurate birth and death dates could be determined since we are aware of his correspondence with Biruni. The number of folios is not known.

۱۸۱ـ غره الزيجات او غره الازياج و معنى زيج بجاى نند.
(اول دو زيج يا زيجها و معنى زيج "ويجى نند")

181. *Ghurrat al-Zijat or Ghurrat al-Azyaj, wa ma'na Zij Bijai Nand.* (The Choicest of the Tables, that Is to Say, the Astronomical Tables of Vijay Nanda).

This is a translation from Sanskrit into Arabic of Vijaya Nanda's book *Karana Tilaka*. It is a valuable book on Indian astronomy. It is in 67 folios. Manuscript: Leiden 1007.

۱۸۲ـ كتاب اللمعات.
(كتاب روشنائيها)

182. *Kitab al-Lamat.* (The Book on Illumination).

This work deals with optics although it is classified under astrology. The number of folios is not known.

۱۸۳ـ ترجمه ما فى پوليسا سدهانتا.
(ترجمه اى در پوليسا سدهانتا)

183. *Tarjamat ma fi Paulisa-Saddhanta.* (Translation of the Contents of *Paulisa-Siddhanta*).

This was a translation from Sanskrit into Arabic of *Paulista Siddhanta*. It also discusses Indian astronomy. The number of folios is not known.

Part III

List of References in English

Articles:

This section includes only articles that appeared in journals. For chapters from books, please see section entitled **Books**.

"Abu Rayhan Beruni." *Variety*, June 18, 1980, p. 24.

Ahmad, N. "Al-Biruni's Geography of India." *Calcutta Geographical Review,* Mar. & Dec. 1943.

Ahmad, N. "Muslim Contribution to Geography During the Middle Ages." *Islamic Culture*, July 1943.

Ahmad, Nazir. "Beruni's *Kitab-as-Saydana* and Its Persian Translation." *Indo-Iranica*, Vol. 14, No. 3, 1961, pp. 5-24.

Ahmad, S. M. et al. "Al-Biruni—An Introduction to His Life and Writings on Indian Sciences." *Indian Journal of History of Science,* Vol. 10, 1975, pp. 98-110; also reproduced in *Afghanistan,* Vol. 24, No. 4, 1972, pp. 54-67.

Ahmad, Z. "Al-Biruni: His Life and His Works." *Islamic Culture*, Vol. 5, July 1931, pp. 343-351.

Ahmad, Z. "Al-Biruni's Research in Trigonometry." *Islamic Culture*, Vol. 6, 1932, pp. 363-369.

"Al-Biruni Millenary Celebrations." *Indo-Iranica,* Vol. 5, No. 4, 1952-53, pp. 14-34.

Barani, S. H. "Al-Biruni's Scientific Achievements." *Indo-Iranica*, Vol. 5, No. 4, 1952-53, pp. 37-48.

Barani, S. H. "Kitabut-Tahdid. (An Unpublished Masterpiece of al-Biruni on Astronomical Geography.)" *Islamic Culture*, Vol. 3, 1957, pp. 165-177.

Beveridge, H. "An Unknown Work of Albiruni." *Journal of the Royal Asiatic Society*, 1902, pp. 333-335.

Bolton, H. C. "The Book of the Balance of Wisdom. An Essay on Determination of Specific Gravity." *American Chemist*, May 1876.

Courtois, V. "Sheikh Abu Raihan al-Biruni." *Islamic Literature*. Vol. 4, Sept. 1952, pp. 35-40.

Dani, A. H. "Al-Biruni on Sanskrit Literature." *Journal of the Pakistan Historical Society*, Vol. 1, 1953, pp. 301-317.

Datta, B. "Al-Biruni and the Origin of the Arabic Numerals." *Proceedings of the Benares Mathematical Society*, Vol. 7, 1925-26, pp. 1-15.

Datta, B. "Two Aryabhatas of al-Biruni." *Bulletin of the Calcutta Mathematical Society*, Vol. 17, Nos. 2-3, 1926, pp. 59-74.

Davidian, M.-L. "Al-Biruni on the Time of Day from Shadow Lengths." *Journal of the American Oriental Society*, Vol. 80, 1960, pp. 330-335.

Derzhavin, V. "Singer, Philosopher and Humanist (Biruni)." *Druzhba Narodov* (Friendship of the People), No. 3, 1965, pp. 233-234.

Frye, R. N. "Notes on the History of Transoxania." *Harvard Journal of Asiatic Studies*, Vol. XIX, 1956, p. 9.

Frye, R. N. and Sayili, A. M. "The Turks in Khurasan and Transoxania at the Time of the Arab Conquest." *The Moslem World*, Vol. XXV, 1945.

Ganguly, D. C. "Al Biruni's Guzerat." *Islamic Culture*, Vol. 13, 1939, pp. 76-78.

Habibi, Abdul Hayee. "The 1000th Birth Anniversary of Allameh Abu Raihan al-Biruni." *Afghanistan*, Vol. 26, No. 1, 1973, pp. 63-67.

Hamarneh, S. K. "The Pharmacy and Materia Medica of al-Biruni and al-Ghafiqi—A Comparison." *Pharmacy in History*, 1976, Vol. 18, No. 1, pp. 3-12.

Haskins, C. "Arabic Science in Western Europe." *Isis*, Vol. XII, 1925, p. 478.

Hill, D. R. "Al-Biruni's Mechanical Calendar." *Annals of Science*, Vol. 42, 1985, pp. 139-163.

Honigmann, E. "The Arabic Translation of Aratus. Phaenomena." *Isis*, Vol. 41, 1950, pp. 30-31.

Hourani, G. "Islamic Cosmology." *Muslim World*, Vol. 56, 1966, pp. 18-22.

Hozayen, A. "Some Arab Contributions to Geography." *Geography Magazine,* Oct. 1932.

Irisov, A. "On the Occasion of the Birth Millenary of Biruni." *Obschestvennye Nauki V. Uzbekistane,* Vol. 1, 1973, pp. 71-72.

Jalali, G. "An Introduction to Biruni." *Afghanistan*, Vol. 27, No. 1, 1974, pp. 47-52.

Kahle, P. "Al-Biruni." *Journal of the Asiatic Society of Pakistan*. Vol. 1, 1956, pp. 18-24.

Karimov, U. I. "On the Date of al-Biruni's Death." *Obschestvennye Nauki V. Uzbekistane*, Vol. 8, 1970.

Karmarkar, R. D. "Hindu Philosophical Literature Known to Alberuni." *Annals of the Bhandarkar Oriental Research Institute*, Vol. 38, 1958, pp. 245-248.

Kazemi, H. A. "Alberuni's Longitudes and Their Conversion into Modern Values." *Islamic Culture*, Vol. XLIX, No. 3, July 1975, pp. 165-176.

Kazemi, H. A. "Al-Beruni on the Shift of the Bed of Amu Darya." *Islamic Culture*, Vol. L, No. 4, Oct. 1976, pp. 201-211.

Kennedy, E. S. "Biruni's Graphical Determination of the Local Meridian." *Scripta Mathematica*, Vol. 24, 1959, pp. 215-255.

Kennedy, E. S. "Al-Biruni on Determining the Meridian." *Mathematics Teacher*, Vol. 56, 1963, pp. 635-637.

Kennedy, E. S., Engle, S., and Vamstad, J. "The Hindu Calendar as Described in al-Biruni's Masudic Canon." *Journal of Near Eastern Studies*, Vol. 24, 1965, pp. 274-284.

Kennedy, E. S. and Muruwwa, A. "Biruni on the Solar Equation." *Journal of Near Eastern Studies*, Vol. 17, 1958, pp. 112-121.

Krachkovsky, I. "Homer and al-Biruni." *Izbrannie Sochineniya* (Selected Works). Moscow-Leningrad. Vol. 2, 1957, pp. 80-87.

Krenkow, F. "Abu'r-Raihan al-Biruni." *Islamic Culture,* Vol. 6, 1932, pp. 528-534.

Krenkow, F. "The Oldest Western Accounts of Chinese Porcelain." *Islamic Culture*, Vol. 7, 1933, pp. 464-471.

Krenkow, F. "The Chapter on Pearls in the Book of Precious Stones by Al-Beruni." *Islamic Culture*, Vol. 15, 1941, pp. 399-421; Vol. 16, 1942, pp. 21-36.

Krenkow, F. "The Drug-Book of Beruni." *Islamic Culture*, Vol. 20, 1946, pp. 109-110.

Law, B. C. "Al-Biruni's Knowledge of Indian Geography." *Indo-Iranica*, Vol. 7, No. 4, 1954, pp. 1-26.

Lesley, M. "Biruni on Rising Times and Daylight Lengths." *Centaurus*, Vol. 5, 1957-58, pp. 121-141.

Levey, M. "Alberuni and Indian Alchemy." *Chymia*, Vol. 7, 1961, pp. 36-39.

Lipkov, A. "Immortal Truths." *Soviet Film*, No. 6, 1975, pp. 21-23.

Lorimer, W. L. "Plato in Afghanistan and India." [Quotation in Biruni] *American Journal of Philology*, Vol. 53, 1932, pp. 157-161.

Majumdar, R. C. "A Passage in Alberuni's India—A Nanda Era?" *Journal of the Bihar and Orissa Research Society*, Vol. 9, 1923, pp. 417-418.

Memon, M. M. "Al-Beruni and His Contribution to Medieval Muslim Geography." *Islamic Culture*, Vol. 33, No. 1, 1959, pp. 213-218.

Meyerhof, M. "A Sketch of Arabic Science." *Journal of the Egyptian Medical Association*, Vol. 19, 1936, pp. 462-474.

Meyerhof, M. "The Article on Aconite from al-Biruni's *Kitab as-Saidana*." *Islamic Culture*, Vol. 19, 1945, pp. 323-328.

Meyerhof, M. "The Earliest Mention of a Manniparous Insect." *Isis*, Vol. 37, 1947, pp. 32-36.

Minorsky, V. "On Sources of Biruni's Informants." *Afghanistan*, Vol. 25, No. 1, 1972, pp. 23-27.

Mishra, J. S. "Al-Biruni's Stay and Travel in India." *Journal of Indian History*, Vol. 44, 1966, pp. 517-528.

Morgenstierne, G. "Modern Indo-Aryan Words in Alberuni's *Indica*." *Indian Linguistics* (*Turner Jubilee Volume*, 1), Vol. 19, 1958, pp. 319-322.

Mukherjee, B. N. "A Critical Examination of Al-Beruni's Charges Against the Copyists of the Puranas." *Indian Historical Quarterly*, Vol. 35, 1959, pp. 132-138.

Najib-ullah. "Abouraihan Al-Beiruni and His Time." *Afghanistan*, Vol. 6, No. 1, 1951, pp. 17-40.

Nazim, M. "A Unique Manuscript on Astronomy." *Annual Report of the Archaeological Survey of India*, 1929-30, pp. 232-233.

Pines, S. "The Semantic Distinction Between the Terms Astronomy and Astrology According to Al-Biruni." *Isis*, Vol. 55, 1964, pp. 343-349.

Rizvi, S. S. H. "A Unique and Unknown Book of al-Beruni: *Ghurrat-uz-Zijat* or *Karana Tilaka*." *Islamic Culture*, Vol. 37, 1963, pp. 112-130, 167-187, 223-245; Vol. 38, 1964, pp. 47-74, 195-212; Vol. 39, 1965, pp. 1-26, 137-180.

Roy Chowdhury, M. L. "Abu Raihan al-Biruni and His Indian Studies." *Indo-Iranica*, Vol. 7, No. 3, 1954, pp. 9-22.

Saidan, A. S. "The Rasa'il of Biruni and Ibn Sinan. A Rearrangement." *Islamic Culture*, Vol. 34, 1960, pp. 173-175.

Sardar, Z. "Islamic Science: Al-Biruni Commemorative Volume." *British Journal for the History of Science*, Vol. 14, No. 48, 1981, pp. 285-286.

Schmidt, O. "The Computation of the Length of the Daylight in Hindu Astronomy." *Isis*, Vol. 35, 1944, pp. 205-211.

Schoy, C. "Moslem Geography of the Middle Ages." *Geographical Review*, Vol. VIV, 1924, pp. 257-269.

Schoy, C. "Al Biruni's Computation of the Value of [Pi]." *American Mathematical Monthly*, Vol. 33, 1926, pp. 323-325.

Schoy, C. "Al-Biruni's Method of Approximation of Chord 40c." *American Mathematical Monthly*, Vol. 33, 1926, pp. 95-96.

Shamsi, F. A. "Abu Rayhan al-Biruni." *Islamic Studies*, Vol. 13, 1974, pp. 179-220.

Sharma, A. "Al-Biruni on the Hindu Notion of Samsara." *Islamic Culture*, July 1977, pp. 165-169.

Siddiqui, A. H. "Muslim Geographic Thought and the Influence of Greek Philosophy." *GeoJournal*, Vol. 37, Sept. 1995, pp. 9-16.

Smith, D. E. and Ginsburg, J. "Rabbi ben Ezra and the Hindu Arabic Problem." *American Mathematical Monthly*, Vol. 25, 1918, pp. 99-108.

Spengler, J. J. "Alberuni: Eleventh-Century Iranian Malthusian?" In *History of Political Economy*, Vol. 3, 1971, pp. 92-104.

Taqizadeh, S. H. "On Biruni's *Chronology*." *Bulletin of the School of Oriental and African Studies*, Vol. 8, 1935-37, p. 947 ff.

Tavadia, J. C. "Al-Biruni and Firdausi." *Visva-bharati Quarterly,* Vol. 15, 1949-50, pp. 51-59.

Taylor, E. G. R. "Some Notes on the Early Ideas of the Form and Size of the Earth." *Geographical Journal*, Vol. LXXXV, Jan. 1935.

Toomer, G. J. "Notes on al-Biruni on Transits." *Orientalia*, Vol. 34, 1965, pp. 45-72.

Usmanov, M. "Ar-Razi and Biruni." *Obschestvennye Nauki v Uzbekistane,* No. 9, 1972, pp. 17-21.

Validi Togan, A. Z. "Islam and the Science of Geography." *Islamic Culture*, Vol. 8, 1934, pp. 511-528.

Validi Togan, A. Z. "Biruni's Picture of the World." *Memoirs of the Archaeological Survey of India*, No. 53, 1937 or 1938.

Wilczynski, J. Z. "On the Presumed Darwinism of Alberuni Eight Hundred Years Before Darwin." *Isis*, Vol. 50, Dec. 1959, pp. 459-466.

Yasin, M. "Social Conditions in Medieval India (As Reflected in the Accounts of al-Biruni, Ibn Battutah, Babur and Bernier), *Quarterly Review of Historical Studies*, Vol. 6, No. 5, 1965, pp. 215-218.

Yasin, M. "Al-Biruni in India." *Islamic Culture,* Vol. L, No. 4, Oct. 1976, pp. 207-213.

Yusuf Ali, A. "Al-Biruni's India." *Islamic Culture*, Vol. 1, 1927, pp. 31-35, 223-230, 473-487.

Zahuri, A. W. "Al-Biruni, a Great Astronomer, Mathematician and Chronologist." *Hamdard Medical Digest*, Vol. 9, Nos. 5-6, 1965, pp. 7-11.

Bibliographies:

Khan, A. S. *A Bibliography of the Works of Abu'l-Raihan al-Biruni.* New Delhi, India: Indian National Academy of Science; Aligarh, India: Centre for West Asian Studies, Aligarh Muslim University, 1982, 77 pages.

This bibliography has the largest number of references by and about Biruni. It includes all available references to materials in all languages published up to 1982. It has a brief introduction in which the author compares Biruni to Ptolemy, Newton, and Einstein. This is the only comprehensive bibliography containing the names of all of Biruni's 183 books.

Nasr, S. H. *al-Biruni: An Annotated Bibliography.* Tehran, Iran: High Council of Culture and Art, Center for Research and Cultural Co-operation. 1973.

This bibliography has 16 brief chapters that are related to the topics under which the references are divided. These chapters are based on Biruni's areas of research. Although this book includes many references in many different languages, Islamic and non-Islamic, the annotations are given only in English and Persian. Most often, one reference appears under several different heading in different chapters.

Books:

This section contains books devoted entirely to Biruni as well as chapters in books. Additionally, we did not hesitate to include books with brief but significant references to our subject. Books were annotated only if available. Chapters in books were not annotated.

Abdullaev, Kh. M. "The Importance of Biruni's Writings in the Development of Science." *Materialy Pervoi Vsesoyuznoi Konferentsii Vostokovedov V.g. Tashkent 4-11, Iyunya*, 1957. (Papers Read at the First All Union Orientalists Scientific Conference in the City of Tashkent, June 4-11, 1957). Tashkent, Uzbekistan: 1958, pp. 115-124.

"Abu Rayhan Biruni: Biruni and His Place in the History of Science of Central Asia." *XXIV Sezd KPSS i Zadacni Razvitiya Obschestvennyx nauk* (Twenty-Fourth Congress of the CPSU [Communist Party of Soviet Union] and Problems of the Growth of Social Sciences). Tashkent, Uzbekistan: "FAN," 1972, pp. 469-476.

"Abu Rayhan Biruni: Biruni and Humanities." *Nauch Sessiya AN UzSSR, Posvyashchennoi 998-i Godovsschine so dnya rozhliniya Abu Raikhana Beruni* (Scientific Session, Academy of Sciences of UzbekSSR, Tashkent, on the Occasion of the 998th Birth Anniversary of Abu Rayhan al-Biruni). Tashkent, Uzbekistan: "FAN," 1972, p. 136.

"Abu Rayhan Biruni: Selected Passage." In *Biruni-Velikii Uchenyi Srednevekovya* (Biruni, the Great Scientist of the Middle Ages), Tashkent, Uzbekistan: 1950, pp. 169-246.

"Abu Rayhan Muhammad Ibn Ahmad Biruni." In Prokhorov, A. M. (Ed.), *Bolshoy Soviet Encyclopedia*, 3rd ed., Vol. 3, Moscow: 1970, pp. 394-395. (Cols. 1170-1171).

Ahmad, A. *Studies in Islamic Culture in the Indian Environment.* Oxford, England: Oxford University Press, 1964, pp. 113-115.

Ahmad, N. *Muslim Contribution to Geography.* Lahore, Pakistan: S. Muhammad Ashraf, 1965, pp. 63-86.

An excellent review of Muslim contributions to geography. It has six very good chapters, a bibliography, and five good maps. Biruni's contribution to geography is discussed in chapter 3.

Ahmad, Q., (Ed.), *India by Al-Biruni.* New Delhi, India: National Book Trust, India, 1983.

A reprint of Biruni's *Book of India* originally translated by Edward Sachau. This book has an excellent introduction focusing on Biruni's contributions to a general understanding of many aspects of Indian society and geography. The author has a positive view of the relationship between Biruni and Sultan Mahmud of Ghazna.

Ahmad, S. M. "The Road-System of India According to Al-Biruni." *Papers Presented to All-India Orientalist Congress.* Aligarh, India: 1966.

Alavi, S. M. Z. *Arab Geography in the Ninth and Tenth Centuries.* Aligarh, India: Aligarh Muslim University, 1965, pp. 33, 54, 106-109.

Alavi, S. M. Z. *Geography in the Middle Ages.* Delhi, India: Sterling Publishers, 1966, pp. 53 and 73.

Ali, S. A. "Al-Biruni: the Scholar and the Writer." *Proceedings of the Pakistan History Conference.* Vol. 3, 1953, pp. 243-252.

Ali, S. M. "Al-Biruni." In *Arab Geography: being the translation of Section II of M. Reinaud's Introduction Generale a la Geographie des Orientaux (Geographie d'Aboulfeda, tome I).* Vol. 1, section II, Aligarh, India: Institute of Islamic Studies, 1960, pp. 81-98.

Professor Ali, a professor of geography at the University of Saugar in India, has translated this book from the French into English. It is a very useful basic reference tool for the understanding of Muslim geography. It has sixteen chapters. Chapter 10 is devoted to Biruni's contribution, and is well-researched. The title of this book is not comprehensive: 7 of the 16 geographers covered are Iranians, not Arabs.

Anklesaria, B. T. "The Names of the Achaemenians in Alberoni." *All-India Oriental Conference, XII.* Vol. 4, No. 3, 1943, pp. 692-702.

Arnold, T. W. "The Caesarean Section in an Arabic Manuscript dated 707 A. H." In Arnold, T. W. and Nicholson, R. A. (Eds.) *A Volume of Oriental Studies Presented to Edward G. Browne.* Cambridge, England: The University Press, 1922, pp. 6-7.

Baker, J. N. L. *A History of Geographical Discovery and Exploration.* New York: Cooper Square Publishers, 1967, pp. 58-74.

Baloch, N. A. "An Evaluation of Beruni's Reference to the Turk Rulers of Kabul and Peshawar Region in the Light of Historical Perspective of the Turkish States and Principalities During the 7th-9th Centuries A.D." In *Acts of the International Symposium on Ibn Turk, Khwarezmi, Farabi, Beyruni and Ibn Sina* (Ankara, 9-12 September 1985). Ankara, Turkey: Ataturk Culture Center, 1990, pp. 23-32.

Barani, S. H. "Muslim Researches in Geodesy." In *Al-Biruni Commemoration Volume.* Calcutta, India: Iran Society, 1951, pp. 1-52.

Barani, S. H. *Al-Biruni and His Magnum Opus al-Qanunu' l-Mas'udi.* Hyderabad Deccan, India: 1956.

Barani, S. H. "Ibn Sina and Alberuni." In *Avicenna Commemoration Volume.* Calcutta, India: Iran Society, 1956, pp. 3-14.

Barthold, W. (Edited by Bosworth, C.), *Turkestan Down to the Mongol Invasion.* London: Luzac, 1968.

Chapters 1 and 2 focus on the geography of Transoxania and Central Asia, where Biruni spent most of his life. This standard work is one of the best and oldest references about Biruni's country.

Beazley, C. R. *The Dawn of Modern Geography.* New York: Peter Smith, Publisher, three volumes, 1949, pp. 530-541.

Belenitzkii, A. M. "A Short Sketch of the Life and Works of Biruni." In *Al-Biruni Sobranie Svedenii dlya poznaniya dragotsennostei (Mineralogiya).* (Al-Biruni, Collected Data Concerning the Knowledge about Precious Stones, Mineralogy). Leningrad, Russia: 1963, pp. 271-291.

Biruni International Congress, Karachi. *Millenary of Abu Raihan Muhammad ibn Ahmad al-Biruni: Papers Presented on the Occasion of al-Biruni International Congress, November 26, 1973 thru December 12, 1973, Pakistan.* Karachi, Pakistan: 1973 or 1974.

Biruni International Congress, Teheran. *The Commemoration Volume of Bir-uni International Congress in Tehran.* Tehran: High Council of Culture and Art, Centre for Research and Cultural Co-ordination, 1976-

Boilot, D. J. "Al-Biruni." In *Encyclopaedia of Islam.* Leiden, Netherlands: E. J. Brill, Vol. 1, 1979, pp. 1236-1238.

Bosworth, C. E. "The Early Ghaznavids." In Frye, R. N. (Ed.), *The Cambridge History of Iran: The Period From the Arab Invasion to the Saljuqs.* New York: Cambridge University Press, Vol. 4, 1975, pp. 162-197.

Bosworth, C. E. "The Ghaznavids." In Asimv, M. S. and Bosworth, C. E. (Edrs.) *History of Civilizations of Central Asia, Volume IV, The Age of Achievement: A. D. 750 to the End of the Fifteenth Century.* Paris: UNESCO Publishing, 1998, pp. 95-117.

Brockelmann, C. "Al-Biruni." In *Encyclopaedia of Islam.* Leiden, Netherlands: E. J. Brill, 1913, Vol. 1, pp. 726-727.

Browne, E. G. *A Literary History of Persia.* London: T. F. Unwin, four volumes, 1902-1906. Also, Cambridge: Cambridge University Press, Four Volumes, 1953-1956, Vol. 1, pp. 158-159, 308-309, 318-319; Vol. 2, pp. 96-98.

As a literary history, this book is not about geographers or scientists. However, Biruni wrote poetry and fiction and is thus represented in this volume. Browne has retold the amazing tale of Biruni's encounter with Sultan Mahmud in which Biruni's punishment follows on the heels of his accurate prediction.

Bulcke, C. "Alberuni and the Rama-Katha." In *Al-Biruni Commemoration Volume.* Calcutta, India: Iran Society, 1951, pp. 71-81.

Carra de Vaux, B. "Baruhima, Brahmans." In *Encyclopaedia of Islam.* Leiden, Netherlands: E. J. Brill, 1913, Vol. 1, p. 726.

Carra de Vaux, B. "Astronomy and Mathematics." In *Legacy of Islam.* Oxford: The Clarendon Press, 1931, pp. 376-397.

Chatterji, S. K. "Al-Biruni and Sanskrit." In *Al-Biruni Commemoration Volume.* Calcutta, India: Iran Society, 1951, pp. 83-100.

Chelkowski, P. (Ed.) *The Scholar and the Saint: Studies in Commemoration of Abul-Rayhan al-Biruni and Jalal al-Din al-Rumi.* New York: Hagop Kevorkian Center for Near Eastern Studies, New York University Press, 1975.

This book has eight chapters by eight primarily Western researchers. It was published by New York University in commemoration of Biruni and Rumi. These articles on a variety of subjects provide an excellent background on Biruni.

Courtois, V. *Al-Biruni; A Life Sketch.* (Indo-Iranica Popular Series, No. 1). Calcutta, India: Iran Society, 1952, 42 pages.

This booklet has four main chapters about Biruni's birthplace, life and character, work and achievements, and his relationship with India. It was presented to the public to mark the occasion of the "Biruni Day" held by the Iran Society in March 1952 to commemorate the millenary of the great scholar's birth.

Dani, A. H. *Alberuni's Indica: A Record of the Cultural History of South Asia about A.D. 1030.* Islamabad, Pakistan: University of Islamabad Press, 1973.

This book is based on Dr. Sachau's translation of Biruni's book on India. It has eleven chapters dealing with the historical geography of Southern India. These chapters cover a wide range of topics: different regions in Southern India; Sanskrit literature; Hindu philosophy, religion, society, eras, and metrology; concepts of Brahmanda, Earth and Heaven; religious views on tides, Linga, and Somanatha.

Davar, F. C. *Iran and India through the Ages*. Bombay: Asia Publishers, 1962.

Dowson, John. "Tarikhu-l Hind of Abu Rihan Al Biruni." In Dowson, John. *The History of India as Told by Its Own Historians*. London: Trubner and Co., 1869. Vol. II, pp. 1-13.

Durant, W. *The Story of Civilization: Part IV, The Age of Faith*. New York: Simon and Schuster, 1950.

Elgood, C. "Persian Science." In *Legacy of Persia*. Oxford: Clarendon Press, 1953, pp. 292-317.

Embree, A. T. (Ed.), *Alberuni's India*. New York: Norton & Company, 1971.

An abridged edition of Biruni's *Book of India* originally translated by Edward Sachau. This volume also has an excellent introduction to Biruni's general work, especially to his research in India.

Embree, A. T. "Foreign Interpreters of India: the Case of al-Biruni." In Chelkowski, P. (Ed.) *The Scholar and the Saint: Studies in Commemoration of Abul-Rayhan al-Biruni and Jalal al-Din al-Rumi*. New York: Hagop Kevorkian Center for Near Eastern Studies, New York University Press, 1975, pp. 1-16.

Esin, E. "Two Cultural Centres in Farabi's Birthland, Kengu-Tarban(d) and Sayram." In *Acts of the International Symposium on Ibn Turk, Khwarezmi, Farabi, Beyruni and Ibn Sina* (Ankara, 9-12 September 1985). Ankara: Ataturk Culture Center, 1990, pp.33-43.

Faul, H. and Faul, C. *It Began with a Stone: A History of Geology from the Stone Age to the Age of Plate Tectonics*. New York: John Wiley & Sons, 1983, pp. 17-21.

Frye, R. N. *The Golden Age of Persia: The Arabs in the East*. New York: Barnes & Noble, 1975, pp. 43-44, 162-163.

Ghirshman, R., Minorsky, V., and Sanghvi, R. *Persia: The Immortal Kingdom*. New York: New York Graphic Society, Orient Commerce Establishment, 1971, pp. 100-119.

Gibb, H. A. R. *The Arab Conquests in Central Asia*. New York: AMS Press, 1970.

This book was originally published in 1923. It provides the reader with a very good brief history and geography of Biruni's homeland before and after the Arab conquests.

Gonda, J. "Remarks on al-Biruni's Quotations from Sanskrit Texts." In *Al-Biruni Commemoration Volume.* Calcutta, India: Iran Society, 1951, pp. 111-118.

Hamarneh, S. K. (Ed.), *Al-Biruni's Book on Pharmacy and Materia Medica: Introduction, Commentary, and Evaluation.* Karachi, Pakistan: Hamdard National Foundation, 1973.

Dr. Sami K. Hamarneh of the Smithsonian Institution wrote this detailed commentary on Said's translation of *Biruni's Book on Pharmacy and Materia Medica.*

Hambly, G. (Ed.), *Central Asia.* New York: Delacorte Press, 1969, Delacorte World History, Volume XVI, pp. 44 and 61.

Hartner, W. and Schramm, M. "Al-Biruni and the Theory of the Solar Apogee: an Example of Originality in Arabic Science." In Crombie, A. C. (Ed.), *Scientific Change.* New York: Basic Books, 1963, pp. 206-218.

Hayes, J. R. (Ed.), *The Genius of Arab Civilization: Source of Renaissance.* New York: New York University Press, 1975, pp. 62, 121, 128, 130, 135, and 162.

Heinan, A. M. "*A Priori* Positions Determining al-Biruni's Scientific Work." In Chelkowski, P. (Ed.) *The Scholar and the Saint: Studies in Commemoration of Abul-Rayhan al-Biruni and Jalal al-Din al-Rumi.* New York: Hagop Kevorkian Center for Near Eastern Studies, New York University Press, 1975, pp. 49-66.

Heras, H. "The Advaita Doctrine in Alberuni." In *Al-Biruni Commemoration Volume.* Calcutta, India: Iran Society, 1951, pp. 116-123; also in *Islamic Literature,* Vol. 11, Mar.-Apr. 1959, pp. 57-60.

Hitti, P. K. *History of the Arabs.* New York: St. Martin's Press, 1970, pp. 376-377.

Hourani, A. *A History of the Arab People.* Cambridge, Mass.: The Belknap Press of Harvard University Press, 1991, pp. 54 and 58.

Iran Society of Calcutta. *Al-Biruni Commemoration Volume.* Calcutta, India: Iran Society, 1951.

This is another commemorative volume published on the occasion of the millenary celebration of Biruni's birth held in Calcutta, India. This commemorative volume contains 21 scholarly papers about Biruni.

Jeffrey, A. "Al-Biruni's Contribution to Comparative Religion." In *Al-Biruni Commemoration Volume.* Calcutta, India: Iran Society, 1951, pp. 125-160.

Kazim, M. A. "Al-Biruni and Trigonometry." In Al-*Biruni Commemoration Volume.* Calcutta, India: Iran Society, 1951, pp. 161-170.

Kazmi, S. H. A. *The Makers of Medieval Muslim Geography: Alberuni.* Delhi, India: Renaissance, 1995.

This book, written in eight chapters, deals with Biruni's life and work as well as with sources on India, regional geography of India, regional geography of Sind, and physical, mathematical and descriptive geographies of India.

Kedrov, B. M., and B. A. Rozenfeld. *Abu Rayhan Biruni.* Moscow: 1973.

Kennedy, E. S. "Biruni." In Gillispie, C. C. (Ed.), *Dictionary of Scientific Biography.* New York: Charles Scribner's Sons, 1970, pp. 147-158.

Kennedy, E. S. *A Commentary on Biruni's Kitab Tahdid al-Amakin: An 11th Century Treatise on Mathematical Geography.* Beirut, Lebanon: American University of Beirut, 1973.

This book, consisting of eleven major chapters, is a highly technical commentary by Professor Kennedy on Biruni's book of determination of places. Professor Kennedy is one of the most knowledgeable Westerners about Biruni's scholarly writings.

Kennedy, E. S. "Al-Biruni on the Muslim Times of Prayer." In Chelkowski, P. (Ed.) *The Scholar and the Saint: Studies in Commemoration of Abul-Rayhan al-Biruni and Jalal al-Din al-Rumi.* New York: Hagop Kevorkian Center for Near Eastern Studies, New York University Press, 1975, pp. 83-94.

Kennedy, E. S. "The Exact Sciences." In Frye, R. N. (Ed.), *The Cambridge History of Iran: The Period From the Arab Invasion to the Saljuqs.* New York: Cambridge University Press, Vol. 4, 1975, pp. 378-395.

Kennedy, E. S. (Translator and commentator.) *The Exhaustive Treatise on Shadows by Abu al-Rayhan Muhammad b. Ahmad al-Biruni.* Aleppo, Syria: Institute for the History of Arabic Science, University of Aleppo, 1976, two volumes.

This is a very impressive translation of Biruni's highly technical book on shadows. Each volume contains thirty chapters, covering such topics as "On the Explanation of Light and Darkness, Luminosity and Shadows" and "Celestial Distances Involving Shadows."

Khan, A. R. "On the Minor Tracts of Abu-Raihan Muhammad bin Ahmad al-Beruni." In *Al-Biruni Commemoration Volume.* Calcutta, India: Iran Society, 1951, pp. 171-175.

Khan, M. "Iran's Contributions to Science and Philosophy." In *Sabade-Gul (Basket of Flower).* Lahore, Pakistan: Nisar Press, 1971, pp. 11-19.

Kimble, G. H. T. *Geography in the Middle Ages*. London: Methuen, 1938.

Kish, G. (Ed.), *A Source Book in Geography*. Cambridge: Harvard University Press, 1978, pp. 211-216.

Kramers, J. H. "Al-Biruni's Determination of Geographical Longitude by Measuring the Distances." In *Al-Biruni Commemoration Volume*. Calcutta, India: Iran Society, 1951, pp. 177-193; also *Analecta orientalia*, 1954, pp. 205-222.

Krenkow, F. "Beruni and the MS. Sultan Fatih No. 3386." In *Al-Biruni Commemoration Volume*. Calcutta, India: Iran Society, 1951, pp. 195-208.

Lawrence, B. B. "The Use of Hindu Religious Texts in al-Biruni's *India* with Special Reference to Patanjali's Yoga-Sutras." In Chelkowski, P. (Ed.) *The Scholar and the Saint: Studies in Commemoration of Abul-Rayhan al-Biruni and Jalal al-Din al-Rumi*. New York: Hagop Kevorkian Center for Near Eastern Studies, New York University Press, 1975, pp. 29-48.

Lawrence, B. B. "Al-Biruni's Approach to Comparative Study of Indian Culture." In Yarshater, E. (Ed.), *Biruni Symposium*. New York: Columbia University, Iran Center, 1976, pp. 27-47.

Le Strange, G. *Lands of the Eastern Caliphate*. Cambridge: Cambridge University Press, 1905, pp. 446-459.

A highly credible and extensively referenced source of information on geography of the Eastern Muslim world. To our knowledge, this is the only book with reference to the two statues of Buddha in Afghanistan studied by Biruni.

Mason, S. F. *A History of the Sciences*. New York: Collier Books, 1962.

Mercier, R. P. "Geodesy." In Harley, J. B. and Woodward, D. (Eds.), *The History of Cartography: Volume Two, Book One, Cartography in the Traditional Islamic and South Asian Societies*. Chicago: The University of Chicago Press, 1992, pp. 175-188.

Meyerhof, M. "Science and Medicine." In *Legacy of Islam*. Oxford: The Clarendon Press, 1931, pp. 314-355.

Minorsky, V. "On Some of Biruni's Informants." In *Al-Biruni Commemoration Volume*. Calcutta, India: Iran Society, 1951, pp. 233-236.

Muhaghegh, M. "The Explanation of Al-Biruni's Dream by Ghazanfar-I Tabrizi." In *Acts of the International Symposium on Ibn Turk, Khwarezmi, Farabi, Beyruni and Ibn Sina* (Ankara, 9-12 September 1985). Ankara: Ataturk Culture Center, 1990, pp. 49-53.

Nainar, M. H. *The Arab's Knowledge of South India*. Madras: 1942.

Nasr, S. H. "The Life, Works, and Significance of al-Biruni." In *An Introduction to Islamic Cosmological Doctrines*. Cambridge: The Belknap Press of Harvard University Press, 1964, pp. 107-121.

Nasr, S. H. *Abu Rayhan Biruni: Scientist and Scholar Extraordinary*. Tehran, Iran: Ministry of Culture and Arts Ores, 1973.

This is an essay containing only 10 pages written by one of the must knowledgeable modern Muslim scholars. It was written on the occasion of Biruni's millenary celebration in Tehran in 1973.

Nasr, S. H. *Science and Civilization in Islam*. Cambridge, Mass.: Harvard University Press, 1968, pp. 50-51, 113-116, 133-138, 179-180, 231-237.

Nasr, S. H. "Life Sciences, Alchemy and Medicine." In Frye, R. N. (Ed.), *The Cambridge History of Iran: The Period from the Arab Invasion to the Saljuqs*. New York: Cambridge University Press, Vol. 4, 1975, pp. 396-418.

Nasr, S. H. "Philosophy and Cosmology." In Frye, R. N. (Ed.), *The Cambridge History of Iran: The Period from the Arab Invasion to the Saljuqs*. New York: Cambridge University Press, Vol. 4, 1975, pp. 419-463.

Nasr, S. H. *Islamic Science: an Illustrated Study*. World of Islam Festival Publishing Company, 1976.

Nasr, S. H. *An Introduction to Islamic Cosmological Doctrines*. Cambridge: The Belknap Press of Harvard University Press, 1964. Also Albany, New York: University of New York Press, 1993.

This book was written by one of the most knowledgeable scholars of Islamic civilization. It has three parts. Part II of this book, with five chapters, deals with Biruni. These brief chapters provide an excellent perspective on almost everything that a researcher should learn about Biruni's life, work and ideas.

Nasr, S. H. and Mohaghegh, M. *Questions and Answers (Between Ibn Sina and Biruni)*. Kuala Lumpur, Malaysia: International Institute for Islamic Thought and Civilization, 1995.

This book deals with a major debate between Biruni and Ibn Sina. Biruni argues for the inductive approach to scientific research and argues against Ibn Sina and the Aristotelian method of deductive reasoning.

Nazim, M. *The Life and Times of Sultan Mahmud of Ghazna*. New Delhi, India: Munshiram Manoharlal, 2nd Edition, 1971.

This is a very good historical work about the life of Sultan Mahmud, especially his religious expedition into India. This book also portrays the relationship between Biruni and Sultan Mahmud as a positive one based on mutual good intentions.

Neugebauer, O. *The Transmission of Planetary Theories in Ancient and Medieval Astronomy.* New York: Scripta Mathematica, Yeshiva University, 1956.

Neugebauer, O. *A History of Ancient Mathematical Astronomy.* Berlin: Springer-Verlag, 1975.

Nizamuddin, M. "Al-Biruni and His Scientific Achievements." In *XXV International Congress of Orientalists.* Moscow: 1960.

Nizamuddin, M. "Al-Biruni, a Pioneer Muslim Scientist." In *'Arshi Presentation Volume.* New Delhi: Majlis-i-Nazr-i 'Arshi, 1965, pp. 73-91.

Nylander, C. "Al-Biruni and Persepolis." *Commemoration Cyrus, Homage Universal,* Vol. 1, 1974, pp. 137-150.

O'Leary, D. L. *How Greek Science Passed to the Arabs.* London: Routledge and K. Paul, 1948, pp. 104-109.

Peters, F. E. "Science, History and Religion: Some Reflections on the *India* of Abul-Rayhan al-Biruni." In Chelkowski, P. (Ed.) *The Scholar and the Saint: Studies in Commemoration of Abul-Rayhan al-Biruni and Jalal al-Din al-Rumi.* New York: Hagop Kevorkian Center for Near Eastern Studies, New York University Press, 1975, pp. 17-28.

Pingree, D. *The Thousands of Abu Mashar.* London: Warburg Institute, 1968, pp. 5-7, 10-11, 29-31, 37-40, 45-48, 56-57, 59-63, 75-76.

Pingree, D. "Al-Biruni's Knowledge of Sanskrit Astronomical Texts." In Chelkowski, P. (Ed.) *The Scholar and the Saint: Studies in Commemoration of Abul-Rayhan al-Biruni and Jalal al-Din al-Rumi.* New York: Hagop Kevorkian Center for Near Eastern Studies, New York University Press, 1975, pp. 67-81.

Pope, A. U. "Alberuni as a Thinker." In *Al-Biruni Commemoration Volume.* Calcutta, India: Iran Society, 1951, pp. 281-285.

Rahman, A. R. *Muslim Contributions to Science and Culture.* Lahore: 1965.

Ritter, H. "The Translation of the Book of Patanjali by Biruni." In *Le livre de millenaire d'Avicenne.* Vol. 2, 1955, pp. 134-148.

Roemer, H. R. "Research on al-Biruni in Germany." In Chelkowski, P. (Ed.) *The Scholar and the Saint: Studies in Commemoration of Abul-Rayhan al-Biruni and Jalal al-Din al-*

Rumi. New York: Hagop Kevorkian Center for Near Eastern Studies, New York University Press, 1975, pp. 95-102.

Rosenthal, F. "Al-Biruni Between Greece and India." In Yarshater, E. (Ed.), *Biruni Symposium*. New York: Columbia University, Iran Center, 1976, pp. 1-12.

Rypka, J. *History of Iranian Literature*. (Edited by Jahn, K.), Dordrecht, Holland: Reidel, 1968, pp. 149-151.

Sadeghi, A. "Some Comments on Persian Words in Al-Biruni's Arabic Writings." In International Symposium on Ibn Turk, Khwarezmi, Farabi, Beyruni and Ibn Sina , 1985, Ankara, Turkey, *Acts of the International Symposium on Ibn Turk, Khwarezmi, Farabi, Beyruni and Ibn Sina* (Ankara, 9-12 September 1985). Ankara: Ataturk Culture Center, 1990, pp. 55-58.

Saffouri, M. and Ifram, A. (Translators). *Al-Biruni on Transits*. Beirut: American University of Beirut, 1959.

This book is a translation of Biruni's book on transits with a commentary by Professor Kennedy. Biruni's original work deals with such subjects as the order of the planets, planetary motion, elevations, and positions of the moon.

Said, H. M. (Ed. and translator). *Al-Biruni's Book on Pharmacy and Materia Medica: Introduction, Commentary, and Evaluation*. Karachi, Pakistan: Hamdard National Foundation, 1973.

This book is Dr. Said's translation of Biruni's book on pharmacy. Biruni's original work, in four major chapters, was a very valuable source of information for its time, the Middle Ages.

Said, H. M *Al-Biruni Commemorative Volume: Proceedings of the International Congress Held in Pakistan, Nov. 26th through Dec. 12th, 1973*. Karachi, Pakistan: Hamdard Academy, 1979.

This major volume contains 73 papers presented at a conference held in Karachi, Pakistan, to commemorate Biruni's millenary birth year in 1973. These papers were published in different languages including Arabic, English, Farsi, and Urdu. This commemorative volume also includes 64 messages from different countries and individuals. Also, 39 scholars from all over the world attended the conference. This volume is an important source of information on Biruni.

Said, H. M. and Zahid, A. *Al-Biruni: His Time, Life and Works*. Karachi, Pakistan: Hamdard Academy, 1981.

This is the only single-volume work written on Biruni's time, life, and accomplishments. Actually this book is a by-product of the proceedings of a conference held in Karachi,

Pakistan, Nov. 26-Dec. 12, 1973, to commemorate the 1000th anniversary of Biruni's birth. The result of this conference was a major *Commemorative Volume,* edited by Hakim Mohammed Said, a physician. Although adding the input of Dr. Ansar Zahid Khan, a historian, was an advantageous step, the book actually does not introduce anything new about Biruni's time, life, or works. This book, indeed, has more material on topics and scholars other than Biruni.

Said, H. M. *The Book Most Comprehensive in Knowledge on Precious Stones: Al-Beruni's Book on Mineralogy.* Islamabad, Pakistan: Pakistan Hijra Council, 1989.

This translation of Biruni's book on precious stones is considered to be the most complete source of information about minerals in the Middle Ages. This book of about 200 pages contains short descriptions of many known precious stones.

Sarton, George. *An Introduction to the History of Science.* Baltimore: The Carnegie Institute, five volumes, 1927-1948, pp. 693-758.

This important book is written in five volumes. In volume 1, Sarton has devoted 18 pages to describing the work and scientific research of Biruni. Sarton was the first to call the first half of the 11th century "the Time of Al-Biruni," signifying the importance of Biruni in the annals of mankind.

Savage-Smith, E. "Celestial Mapping." In Harley, J. B and Woodward, D. (Eds.), *The History of Cartography: Volume Two, Book One, Cartography in the Traditional Islamic and South Asian Societies.* Chicago: The University of Chicago Press, 1992, pp. 12-70.

Sayili, A. *The Observatory in Islam.* Ankara: Türk Tarikh Kurumu Basimevi, 1960, pp. 56-61, 91-94, 109-111, 116-120, 124-130, etc.

Sezgin, F. *The Contribution of the Arabic-Islamic Geographers to the Formation of the World Map.* Frankfort, Germany: IGAIW, 1987, pp. 21-27.

Sharif, M. M. (Ed.), *A History of Muslim Philosophy.* Wiesbaden, Germany: Harrassowitz, two volumes, 1963-1966.

Sharipov, A. *The Great Thinker Biruni.* Tashkent: 1972.

Sharma, Arvind. *Studies in "Alberuni's India."* Wiesbaden, Germany: Harrassowitz, 1983.

This book has 3 chapters composed of 15 short sections on Biruni's observations in India. The three main sections of the book are "Albiruni and the Bhagavadgita," "Albiruni on the Concept of Avatara in Hinduism," and "Albiruni on Hindu Xenophobia."

Soucek, P. "An Illustrated Manuscript of al-Biruni's Chronology of Ancient Nations." In Chelkowski, P. (Ed.) *The Scholar and the Saint: Studies in Commemoration of Abul-*

Rayhan al-Biruni and Jalal al-Din al-Rumi. New York: Hagop Kevorkian Center for Near Eastern Studies, New York University Press, 1975, pp. 103-168.

Spuler, B. "The Disintegration of the Caliphate in the East." *In Holt, P. M., Lambton, A.K. S., and Lewis, B. (Eds.),* The Cambridge History of Islam. *Cambridge: Cambridge University Press, Vol. 1, 1970, pp. 143-161.*

Sykes, P. *A History of Persia*. New York: Barnes & Noble, Inc., 1969, pp. 26-27, 50-56.

Tavadia, J. C. "Al-Biruni and Orientalistics." In *Al-Biruni Commemoration Volume.* Calcutta, India: Iran Society, 1951, pp. 287-291.

Tibbitts, G. R. "Later Cartographic Development." In Harley, J. B. and Woodward, D. (Eds.), *The History of Cartography: Volume Two, Book One, Cartography in the Traditional Islamic and South Asian Societies*. Chicago: The University of Chicago Press, 1992, pp. 137-155.

Togan, A. (Ed.) *Biruni's Picture of the World.* [New Delhi, India?]: 1971. Publication no. 53, *Memoirs of the Archaeological Survey of India.*

Professor Togan culled from a variety of sources written by Biruni to produce a book called *Picture of the World.* This book, which is in Arabic, includes Togan's preface in English.

Tolstov, S. P. "Biruni and His Times." In *Biruni Sbornik.* Moscow-Leningrad: 1950, pp. 3-29.

Turner, H. R. *Science in Medieval Islam*. Austin: University of Texas Press, 1997, pp. 65, 116.

Usmanov, M. "Al-Biruni as a Thinker." In *Izv. AN. USSR ser. Istorii i Filosofii (Bulletin of the Academy of Sciences of Uz SSR—History & Philosophy*). Series 6, No. 2, 1949, pp. 120-133; also in *Biruni Sbornik*, Moscow-Leningrad: 1950, pp. 30-54.

Wescoat, J. L. "Muslim Contributions to Geography and Environmental Ethics: The Challenges of Comparison and Pluralism." In Light, A. and Smith, J. M. (Eds.), *Space, Place, and Environmental Ethics*. New York: Rowman & Littlefield Publishers, Inc., 1977, pp. 91-116.

Winter, H. J. J. *Eastern Science*. (*Wisdom of the East Series*). London: J. Murray, 1952.

Yarshater, E. "List of the Achaemenid Kings in Biruni and Bar Hebraeus." In Yarshater, E. and Bishop, D. (Eds.), *Biruni Symposium*. New York: Columbia University, Iran Center, 1976, pp. 49-65.

Yarshater, E. and Bishop, D. (Eds.), *Biruni Symposium*. New York: Columbia University, Iran Center, 1976, 65 pages. Also published in paperback by Caravan Books, January 1983.

This booklet contains four major papers about Biruni written by Franz Rosenthal of Yale University, G. H. Youssefi of Mashad University, Bruce B. Lawrence of Duke University, and Ehsan Yarshater of Columbia University. The first three papers were presented in a symposium held by the Iran Center at Columbia University in December of 1973 on the occasion of the 1000th anniversary of Biruni's birth. The last paper was read at the Biruni International Congress held in September 1973 in Tehran. Together, they are among the best existing pieces of research on Biruni.

Yarshater, E. (Ed.) "Biruni." In *Encyclopaedia Iranica*, Vol. IV, New York: Routledge & Kegan Paul, 1990, pp. 274- 287.

This is one of the best brief introductions written on Biruni. It has eight sections written by experts on various aspects of Biruni's life and works: Bosworth, Pingree, Saliba, Anawati, and de Blois.

Youssefi, G. H. "Abu-Reyhan Biruni, A Lover of Truth." In Yarshater, E. and Bishop, D. (Eds.), *Biruni Symposium*. New York: Columbia University, Iran Center, 1976, pp. 13-26.

Zahidov, V. Yu. "Al-Biruni, the Great Uzbek, Savant of the Middle Ages." In *Biruni-Velikii Uchenyi Srednevekovya*. Tashkent: Uz SSR, 1950.

Dissertation:

Ghayasuddin, M. *A Critical Analysis of the Writings of Abu'l Rayhan Muhammad al-Biruni, Pertaining to India.* Unpublished Dissertation for Ph. D. degree in Arabic, Aligarh Muslim University, Aligarh, India, 1968.

We have tried several times to get some more information on this dissertation from Aligarh Muslim University. So far, we have not received any positive answer. Actually, on June 21, 2004, Dr Fazal, Chiarman of Geography Department at AMU, sent us an email indicating that as a policy photocopying a "Thesis Abstract" was not allowed at his university. Unfortunately, a "Thesis Abstract" for the above research cannot be located elsewhere.

Film:

Abu Rajchan Biruni (or *Abu Rajkhan Beruni* or *Abu Reikhan Biruni*). Director: Shukhrat Abbasov. Writers: Shukhrat Abbasov and Pavel Bulgakov. Tashkent: Usbekfilm, 1974.

Director Shukhrat Abbasov explores the life and achievements of the great Islamic scholar Abu Raihan al-Biruni in this informative costume drama. Al-Biruni was born in 973 AD in the town of Kheva in present-day Uzbekistan. His brilliant scholarship came to the attention of Sultan Ghaznawi, the ruler who conquered the region, and while he was still young, al-Biruni traveled with the sultan through India. He translated Sanskrit texts into Arabic, recorded the history and culture of India, and developed the fields of astronomy and mathematics, the ancient history of nations, geography, and medicine -- to name a few. He resolved several scientific questions and is still considered to be one of the outstanding thinkers in world history. His achievements are justifiably lauded in this docudrama that also brings up the difficulties he faced with the "establishment" of his day. ~ Eleanor Mannikka, All Movie Guide

Part IV

Selected Internet Sites on Biruni

In addition to traditional sources, we decided to investigate the Internet. We utilized the following five search engines to retrieve documents dealing with Biruni.

Table 3
Search Engines Used for this Bibliography

Search Engines	Results (number of hits)
www.alltheweb.com	4,765
www.altavista.com	3,109
www.google.com	7,150
www.mamma.com	34
www.metacrawler.com	59

As can be seen from the table above, we retrieved a wide range of results, with the lowest number obtained from www.mamma.com and the highest number of results from www.google.com. Looking at all 7,150 items in the latter would have been impractical, but because *Google* gave the most comprehensive results, we were interested in further research using this search engine. Thus we constructed an "Advanced Search" using two limits: English language materials only, and a required appearance of Biruni's name in the title of each result. This search procedure resulted in a retrieval of nearly 100 references. After further examination of these results, eliminating duplicate entries and false hits, it was determined that 39 items would be included.

The following section contains the names of these 39 sites. The style of the entries below follows the format for electronic sources set forth in the American Psychological Association's *Manual of Style*, 5th edition. To this we added our own brief evaluation indicating how much information each site provides and how useful the site is.

It is our opinion that of the following 39 results, very few can be considered useful in research. Several are duplicates available at multiple Internet addresses. Of the remaining entries, few provide in-depth and comprehensive coverage of Biruni's scholarly research. The article written by J. J. O'Connor and E. F. Robertson of the School of Mathematics and Statistics, University of St. Andrews, Scotland, was one of the best sources we found. Another excellent source of books on and by Biruni was the site of the Library of Congress. However, it is limited due to the fact that the books it lists are drawn solely from the Library's collection.

The quality of the resources that we investigated on the Internet as of June 2004 was not impressive and was, in our opinion, disappointing. Internet sources on Biruni are overloaded with repetitious, error-filled material and "infomercials," obscuring the presence of any useful information. At present, in a search for specific topics like that of Biruni's life and work, a traditional library search is recommended. However, it is to be hoped that future developments in technology and scholarship may eventually result in a more valuable assortment of information about Biruni on the Internet.

1. Aber, J. S. *Abu Rayhan al-Biruni: History of Geology.* (2003). Retrieved May 27, 2004, from Emporia State University Web site: http://academic.emporia.edu/aberjame/histgeol/biruni/biruni.htm (10k)

   Written by James S. Aber, this colorful site discusses Biruni's contribution to geology and provides links to related Web sites.

2. *Abu Arrayhan Muhammad Ibn Ahmad Al-Biruni.* (n.d.). Retrieved May 26, 2004, from http://www.zia.rahin.iwarp.com/albiruni.htm (34k)

   This site contains the same article as entry 1 above, unattributed, at the stated Internet address.

3. *Abu Raihan al-Biruni (973-1048): Library of Congress citations.* (n.d.). Retrieved May 30, 2004, from http://www.mala.bc.ca/~mcneil/cit/citlcbiruni1.htm (29k)

   Bibliographic records for twenty works by and about Biruni in the collection of the Library of Congress. Most of the works are in English. Also includes nine records that establish bibliographic headings related to Biruni, serving to unify various forms of Biruni's name and names of his works.

4. *Abu Raihan al-Biruni (973—1048 A.D.). (n.d.).* Retrieved May 26, 2004, from http://members.tripod.com/~wzzz/BIRUNI.html (10k)

   This brief, unattributed document deals with only a few of Biruni's books and offers little detailed information. See entry 12 for attribution.

5. *Abu Raihan al-Biruni (973—1048 A.D.).* (n.d.). Retrieved May 27, 2004, from http://islamicity.com/Science/Scientists/Al-Biruni.htm (6k)

   This site contains the same unattributed document as entry 5 above, at the stated Internet address. See entry 12 for attribution.

6. *Abu Raihan al-Biruni (973—1048 A.D.).* (n.d.). Retrieved May 27, 2004, from http://ahram.org/islamic/albiruni.htm (6k)

   This site contains the same unattributed document as entries 5 and 6 above, at the stated Internet address. See entry 12 for attribution.

7. *Abu Raihan al-Biruni (973—1048 A.D.).* (n.d.). In *Muslim Thinkers and Scientists.* Retrieved June 15, 2004, from http://www.iman.co.nz/index.htm?innerpage_think.htm?think/al_biruni.htm

This site contains a brief biography of Biruni and the names of some of his books.

8. *Abu Raihan Biruni, philosopher, physicist, mathematician and astronomer.* (n.d.). In *Iranian personalities.* Retrieved May 27, 2004 from Iran Chamber Society Web site: http://www.iranchamber.com/personalities/abiruni/abu_biruni.php (15k)

This site contains the same unattributed document in entries 5-7 above, at the stated Internet address. See entry 12 for attribution.

9. *Abu-Rayhaan-e-Birooni.* (n.d.). Retrieved June 15, 2004, from http://www.farhangsara.com/biruni.htm (14k)

This brief biography of Biruni, available in both Persian and English, provides an excellent brief introduction to most of Biruni's most famous books.

10. *Abu Rayhan Muhammad ibn Ahmad Al-Biruni.* (2003). In *The Columbia electronic encyclopedia.* Retrieved May 30, 2004 from http://reference.allrefer.com/encyclopedia/A/AlBiruni.html (37k)

Another very brief encyclopedia article with general information on Biruni's life and work.

11. Al-Biruni. (n.d.). *Albalagh*, 2004, April 17. Retrieved May 27, 2004, from http://www.albalagh.net/kids/history/biruni.shtml (9k)

Very brief general introduction to Biruni's life and achievements. From the children's section of the e-journal/e-newsletter *Albalagh.*

12. *Al-Biruni.* (n.d.). Retrieved June 15, 2004, from http://www.smartpedia.com/smart/browse/Al-Biruni
This brief biography of Biruni lists his scholarly research activities and the titles of some of his books.

13. *Al-Biruni.* (2003, April 13). Retrieved June 15, 2003, from http://www.languagehat.com/archives/000535.php (11k)

A brief introduction to Biruni's life, this article expresses the opinion that more people should be aware of Biruni.

14. *Al-Biruni.* (2004, April 7). In *Wikipedia: The Free Encyclopedia.* Retrieved May 27, 2004, from www.wikipedia.org/wiki/Biruni (10k)

This site also provides minimal information on Biruni. Please note that *Wikipedia* is a "free content encyclopedia" which means that the content can be edited by anyone who clicks on the "edit this page" link.

15. *Al-Biruni (973 – 1048).* (2001-2002). Retrieved May 30, 2004, at ForumOnAstrology's Web site: http://www.forumonastrology.com/astrologers/al-biruni.html (2k)

This document, located on the site of a company that offers an online course in astrology, contains Biruni's natal chart with astrological symbols.

16. *Al-Biruni, 973 – 1048, Arab scholar.* (n.d.). Retrieved May 27, 2004, from: http://www.hyperhistory.com/online_n2/people_n2/persons5_n2/albiruni.html (2k)

This thumbnail biography of Biruni provides minimal information.

17. *Al-Biruni, Abu Rayhan Muhammad ibn Ahmad.* (2001). In *The Columbia encyclopedia* (6th ed.). New York: Columbia University Press. Retrieved May 27, 2004, from Project Bartleby Web site: www.bartleby.com/65/al/AlBiruni.html (19k)

Brief article from the *Columbia Encyclopedia*, 6th edition, 2001.

18. *Al-Biruni or Al Beruni, Abu Rayhan Muhammad ibn Ahmad.* (2003). In *The Columbia electronic encyclopedia.* (6th ed.). Retrieved May 27, 2004, from www.factmonster.com/ce6/people/A0803100.html (22k)

This site contains the same article as entry 23 above, at the stated Internet address.

19. *Al-Biruni or Al Beruni, Abu Rayhan Muhammad ibn Ahmad.* (2003). In *The Columbia electronic encyclopedia.* (6th ed.). Retrieved May 27, 2004, from www.infoplease.com/ce5/CE001189.html (28k)

A brief article about Biruni from the 2003 edition of the *Columbia Electronic Encyclopedia*.

20. *Al-Biruni world map of the distribution of land and sea, 1029 A.D.* (n.d.). Retrieved May 27, 2004, from http://gate.henry-davis.com/MAPS/EMwebpages/214.3.html (2k)

Annotated version of Biruni's world map of the distribution of land and sea, 1029 C.E. (9.5 cm. diameter), oriented with the South at the top.

21. Argis, A. (2003, June 8). *Ahmad al-Biruni.* Retrieved June 15, 2004, from http://www.newsfinder.org/more.php?id=774_0_1_0_M (46k)

This two-page article was written by Ali Argis in June of 2003. It is interesting that in the midst of several paragraphs of general information about Biruni Argis presents Biruni's precise calculation of the power of doubling.

22. *Biruni.* (n.d.). Retrieved May 28, 2004, from www.irangasht.co.uk/sci-Biruni.htm (30k)

A brief description of Biruni's life and works maintained by Irangasht.com, a site whose mission is to promote Iranian culture, tourism and industry.

23. *Biruni.* (1911). In *Encyclopedia Britannica.* (11th ed.). Retrieved May 27, 2004, from http://85.1911encyclopedia.org/B/BI/BIRUNI.htm (7k)

This very brief encyclopedia entry unfortunately contains a number of inaccuracies.

24. [Biruni]. (2004, April 1). Retrieved May 30, 2004, from Demir Group Web site: http://www.demirgroup.uz/ENG/E05.htm (3k)

This short paragraph (in awkward English) about Biruni's research is one section of an article on Uzbekistan, and is illustrated with beautiful, full-color pictures.

25. *Chronology of the Muslim scholar Al-Biruni (973 - 1048).* (n.d.). Retrieved June 1, 2004, from http://www.holyfire.org/eng/doc_BiruniIntheGarden.htm (8k)

This single page appears to be a translation of a passage from the 1991 book *In the Garden of Science*, published by Reclam. This passage is based on Biruni's *Chronology of Ancient Nations* and contains only a brief reference to some Christian holidays.

26. *A History of Muslim pharmacy: Al-Biruni and al-Ghafiqi in the balance.* (n.d.). Retrieved May 27, 2004, from the Islamic Organization for Medical Sciences Web site: http://www.islamset.com/heritage/pharmacy/balance.html (6k)

Compares the science of pharmacy as presented by Biruni and Ghafiqi from Andalus, today's Spain.

27. *A History of Muslim pharmacy: Al-Biruni and his work.* (n.d.). Retrieved May 27, 2004, from the Islamic Organization for Medical Sciences Web site: http://www.islamset.com/heritage/pharmacy/balance.html (6k)

Focuses solely on Biruni's book of pharmacy titled *Saydanah.*

28. *Islamic Scholars: Abu Raihan Biruni.* (2003, January 23). Retrieved May 28, 2004, from www.islamicdigest.net/id5/article.php?sid=1290 (52k)

Although it contains some inaccuracies and resides on a site with controversial content, this unattributed document provides useful general information on Biruni's life and research.

29. Manugian, M. (1980?). [*Abu Rayhan Biruni, A Scholar Who Lived a Thousand Years Ago with Ideas for the 20th Century*]. Retrieved June 15, 2004, from http://www.azargoshnasp.net/~iran/famous/biruni_khwarazmi/birunikhwarazmimain.htm (3k)

This nine page scanned document is a valuable article in Persian. The title provided above is a translation of the Persian title.

30. Marvin, C. *Al-Biruni.* (1995-2000). In *The Window: Philosophy on the Internet: Philosophers.* Retrieved May 27, 2004 from Trinity College, Philosophy Department Web site: http://www.trincoll.edu/depts/phil/philo/phils/muslim/biruni.html (8k)

This is the original article, by Christopher A. Marvin, that appears unattributed in entries 5-7 above. It is a brief article, deals with only a few of Biruni's books, and offers little detail

31. McNeil, R. (1995-2004). *Abu Raihan al-Biruni.* In *Malaspina great books.* Retrieved May 27, 2004, from http://www.mala.bc.ca/~mcneil/biruni1.htm (16k)

This site provides a great number of links to citations of Biruni's books appearing in other Internet sites.

32. *Medieval sourcebook: Al Biruni (973-1048 CE): The existing monuments or chronology, c. 1030 CE.* (1917). Retrieved May 27, 2004, from the Fordham University Center for Medieval Studies Web site: www.fordham.edu/halsall/source/1030al-biruni1.html

This site, copyrighted by Paul Halsall in August, 1998, is a part of the *Internet medieval source book*, a collection of public domain and copy-permitted texts related to medieval and Byzantine history. The site contains Prof. Arkenberg's modernized text of an article on Biruni taken from: Horne, C. (Ed.). (1917). T*he Sacred books and early literature of the East:* Vol. 6. *Medieval Arabia* (pp. 92-96). New York: Parke, Austin & Lipsomb. This site looks exclusively at Biruni's book of *Chronology of ancient nations.*

33. O'Connor, J. J., Robertson, E. F. (1999). *Abu Arrayhan Muhammad ibn Ahmad al-Biruni.* In *The MacTutor History of Mathematics Archive.* Retrieved March 30, 2004, from University of St. Andrews, Scotland, School of Mathematics and Statistics Web site: http://www-history.mcs.st-andrews.ac.uk/Mathematicians/Al-Biruni.html (27k)

This is a very informative seven-page essay on the life and works of Biruni, written by John J. O'Connor and Edmund F. Robertson of the School of Mathematics and Statistics, University of St. Andrews, Scotland. It includes several links to glossary entries, a list of 72 references, some quotations, a poster, mathematicians born in the same country as Biruni, chronology, honors awarded to Biruni, and more. This site is highly recommended for general information.

34. *Personalities in Islam: Al-Biruni (973-1048 AD).* (n.d.). Retrieved May 30, 2004, from Islamic Paths Web site: http://www.islamic-paths.org/Home/English/History/Personalities/Content/Biruni.htm (36k)

This site provides general information about Biruni's life and research, and also describes some of his important books.

35. *U.S. Embassy Presents Abu Raykhan Biruni Institute of Oriental Studies with Cultural Preservation Grant.* (2001, September 26). Retrieved May 27, 2004, from the Embassy of the United States, Tashkent, Uzbekistan Web site: http://www.usembassy.uz/home/index.aspx?&mid=217&lid=1&overview=535 (5k)

Announcement of a grant from the U.S. Embassy in the Republic of Uzbekistan to assist with microfilming, photocopying, and restoring Biruni's peerless collection of manuscripts.

36. Warnock, C. *Al-Biruni.* (2001). Retrieved May 27, 2004, from www.renaissanceastrology.com/albiruni.html (12k)

This site deals primarily with Biruni's contribution to astrology, including his horoscope, with many astrological symbols illustrated.

37. *Weather: Biruni, Uzbekistan..* (2004, May 28). Retrieved May 28, 2004, from http://weather.lycos.com/?q=Biruni%2C+UZ (38k)

This site provides daily weather reports for the city of Biruni in the Republic of Uzbekistan.

38. Zahoor, A. (2002). *Abu Raihan Muhammad al-Biruni (973 – 1048 CE).* Retrieved May 26, 2004, from http://www.cyberistan.org/islamic/biruni.html (3k)

This site is actually an advertisement for an e-book, *Islamic Civilization,* which can be ordered through the site for $65.00. ISBN 0-9702389-3-2. This e-book contains over 115 articles on Muslim society and its contributions beginning with the eighth century, including a biography of Biruni..

39. Zahoor, A. (1997). *Abu Raihan Muhammad al-Biruni (973-1048 C.E.).* In *Muslim scientists, 700 - 1500 C.E. (before European Renaissance).* Retrieved May 26, 2004, from http://www.unhas.ac.id/~rhiza/saintis/biruni.html (10k)

Dr. A. Zahoor's brief article outlines Biruni's contributions to science.

Part V

Persian References

This part presents books, articles, and bibliographies that are written in Persian. In this part we followed the same format that we discussed in the previous section. For this part, we found eleven books in Persian that were written on Biruni. The great majority of these books, a total of eight, were written for the occasion of the 1000th anniversary of Biruni's birth. We also identified four books that have chapters devoted to Biruni, namely: *Famous Muslim Scientists* by Khayyery (1958); *The Great Iranian Mathematicians* by Qorbani (1971); *History of Philosophy in Islamic Civilization Until the Fifth Century* by Safa (1957); and *History of the Chosen Ones* by Sepehrom (1962).

Next, we found 26 articles on Biruni. Three of these articles appeared in a journal called *Aryana*, published in Kabul, Afghanistan. One of these articles was published in *Al-Biruni Commemoration Volume*, Calcutta, India, by Iran Society in 1951. Another was published in the *Magazine of Oriental College of Lahore*, Pakistan. In addition, two articles by Nafisi (1952) and Safa (1953) both appeared in *Indo-Iranica*, Calcutta, India. Thus, the great majority of these articles, 22 out of 26, were published in Tehran, Iran.

As mentioned before in the section on Methodology, we also found two important bibliographies on Biruni and utilized them in the present research. Again, both of them were prepared for Biruni's birthday millenary. Azkayi's (1973) *Catalogue* was used to verify the titles of Biruni's books. Only a few of our references, particularly articles in Persian, were selected from Nasr's (1973) *Bibliography*.

Persian References:

منابع فارسی

مقـــالات: (Articles)

آندرس، ا. ک. "اثر مفقود شدهٔ ابوریحان بیرونی در بارهٔ خوارزم طبق منابع تاریخ بیهقی." پیام نوین، جلد ۱، ۱۳۳۸_۱۳۳۷، شمارهٔ ۵، صفحات ۹۵_۹۴.

پروین گنابادی، محمد. "آیا ابوریحان با داروین هم عقیده بوده است." سخن ماه، دورهٔ چهارم، ۱۳۳۱، شمارهٔ ۷، صفحهٔ ؟۵٤.

ترکی، محمد قادر و گویا، سرور. "نامعلوم بیرونی." آریانا، کابل: ۵، ۱۳۱۵_۱۳۱٤، صفحات ۹٦_۹۱.

تمدن، محمد. "ابوریحان بیرونی." نامهٔ علمی و فنی، جلد ۲، ۱۳۳۷، شمارهٔ ٤، صفحات ٦_۱.

تهرانی، سید جلال الدین. "تاریخ مختصر نجوم ایرانیان و اعراب." گاهنامه، شمارهٔ ۱، صفحات ۱۳۲_۱۰٤.

تهرانی، سید جلال الدین. "تاریخ ارصاد مهمه." گاهنامه، شمارهٔ ۵، صفحات ۷۶_۳۷.

تهرانی، سید جلال الدین. "آلات رصدیهٔ قدما." گاهنامه، شمارهٔ ۵، صفحات ۱۲٤_۹۷.

تهرانی، سید جلال الدین. "تاریخ مختصری از مراصد اسلامی." گاهنامه، شمارهٔ ۵، صفحات ۲۱۰_۲۰۲.

تهرانی، سید جلال الدین. "اصلاحات علمیهٔ علم احکام نجوم." گاهنامه، شمارهٔ ٦، صفحات ۷٦_٤۹.

حبیبی، عبدالحی. "بیرون و بیرونی." آریانا، کابل: ٦، بخش نهم، ۱۳۲۵_۱۳۲٤، صفحات ۸۸۸_۸۸٤.

حبیبی، عبدالحی. "یک صفحه از حیات بیرونی." آریانا، کابل: ٦، بخش نهم، ۱۳۲۷_۱۳۲٦، صفحات ۳۰_۲۹.

داناسرشت، اکبر. "ابوریحان بیرونی." سخن علمی، شمارهٔ ۵، ۱۳٤۵، صفحات ۳۹۸_۲۹۷.

راهنما، غلامحسین. "ریاضیدانِ برجسته ابوریحان بیرونی." کنجکاویهای علمی و ادبی، صفحات ۵٤_٤٦.

ریتر، هـ. "ابوریحان و کتاب پاتنجلی." یغما، سال اول، ۱۳۳۲، شمارهٔ ۷، صفحات ۱۹_۱۵ و ٦٦_٦۰.

زهیر، ع. ق. "ابوریحان بیرونی." ادب، جلد ۲، شمارهٔ ٤_۳، صفحات ٦۲_٥۲ و جلد ۳، ۱۳۲٤، صفحات ٥۱_٤٥.

سید عبدالله. "لغات هندی در کتاب قدیم عربی مثل آثار ابوریحان." مجلهٔ اورینتال کالج لاهور، شمارهٔ ۱۹، بخش سوم، ۱۹٤۳_۱۹٤۲، صفحات ٤۳_٤۲.

شهریاری، پرویز. "ابوریحان بیرونی." دانش و مردم، شمارهٔ ۱ و ۲ سال سوم، ۱۳۸۱؟، صفحات ۲۸_۳۹.

قوامی، رضی. "نگاهی به کتاب کانه شناسی ابوریحان بیرونی." سخن علمی، شمارهٔ ٤، ۱۳٤٤، صفحات ۹۱_۸۹.

مصاحب، غلامحسین. "خلاصه ای از افکار ریاضی بیرونی و فهرست آثار مهم ریاضی او." جبر و مقابلهٔ خیام. تهران: ۱۳۱۷، صفحات ۱۲۷_۱۳۳.

معین، محمد. "فوائد لغوی کتاب المجاهر بیرونی." ***Al-Biruni Commemoration Volume***, 1951, pp. 237-246.

منظور الحق، "ابوریحان بیرونی." هلال، جلد ۱، ۱۳۳۲_۱۳۳۱، شمارهٔ ٦، صفحات ۳٤_۲۹ و ٥٥.

نصر، سید حسین. "نظریات ابوریحان بیرونی در بارهٔ معنی طبیعت و روشهای تحقیق در علوم طبیعی." مجلهٔ دانشکدهٔ ادبیات، تهران: جلد ۷، ۱۳۲۹_۱۳۲۸، شمارهٔ ٤، صفحات ۹۰_۷۸.

نظامی عروضی سمرقندی، چهار مقاله. تهران: ۱۳۳۳، به تصحیح محمد معین، صفحات گوناگون مخصوصاً مقالت سوم، صفحهٔ ۸۷ به بعد.

نفیسی، سعید. "آثار چاپ شدهٔ ابوریحان بیرونی." ***Indo-Iranica***، کلکته: شمارهٔ ۵، بخش چهارم، صفحات ٤_۱، ۱۹۵۲.

نفیسی، سعید. "ابوریحان بیرونی." پیام نو، جلد شمارهٔ ۱۲_۱۱ صفحات ۲۲_۲٦.

صفا، ذبیح الله. "برخی از نظریات فلسفی ابوریحان و مختصری در مناقشاتِ علمی او و ابن سینا." ***Indo-Iranica***، کلکته: شمارهٔ ۵، بخش چهارم، ۱۹۵۲، صفحات ۱۲_۵.

کتابشناسیها: (Bibliographies)

اذکایی، پرویز. کارنامهِ ی بیرونی (کتابشناسی و فهرست آثار). تهران: وزارت فرهنگ و هنر، مرکز پژوهشهای مردمشناسی و فرهنگ عامه، آبان ماهٔ ۱۳۵۲.

پرویز اذکایی یکی از استادان متخصص ایرانی در مورد بیرونی میباشد. این کتاب دارای هشت فصل در بارهٔ فهرست آثار بیرونی بوده و باید بعنوان کتابشناسی بیرونی به آن نگاه نمود. در فصل سوم اسامی کامل ۱۸۰ جلد کتاب که به نام بیرونی معروف هستند به دو زبان عربی و فارسی و اسامی کتابخانه هایی که دارای این کتابها هستند نقل شده است.

نصر، سید حسین. کتابشناسی توصیفی ابوریحان بیرونی. تهران: شورای عالی فرهنگ و هنر مرکز مطالعات و هماهنگی فرهنگی، شمارهٔ ۸، ۱۳۵۲.

این کتاب اکثر منابع چاپ شده در رابطه با بیرونی را تا سال ۱۳۵۲ شمسی در بر دارد. این منابع به دو قسمت زبانهای اسلامی و غیر اسلامی تقسیم شده اند. بیشتر منابع تحت عنوانهایی مختلف تکرار گردیده اند.

کتابها: (Books)

اذکایی، پرویز. افکار و آراء بیرونی، دانشمند بزرگ ایران. تهران: انتشارات طرح نو، ۱۳۷۴.

پنج مقاله در شرح حال و آثار ابوریحان بیرونی: بمناسبت هزارمین سال تولد نابغهٔ بزرگ ایران. کرمان: ادارهٔ کل فرهنگ و هنر استان کرمان، شهریور ۱۳۵۲.

این کتاب دارای پنج مقالهٔ کوتاه میباشد که بوسیلهٔ دکتر ذبیح الله صفا، دکتر سید جعفر سجادی، دکتر سعید فاطمی، قدمعلی سرام و بهروز صوراسرافیل نوشته شده اند.

حکیمیان، ابوالفتح. زندگینامهٔ ابوریحان. تهران: چاپ آفتاب، ۱۳۵۲.

این کتاب در سیزده فصل بوسیلهٔ به رشتهٔ تحریر درآمده است. قسمت اول این کتاب، پنج فصل تاریخچهٔ زندگی پُر ماجرای بیرونی را در بر میگیرد. بقیهٔ کتاب راجع به آثار او میباشد.

خیری، سید محمود. دانشمندان نامی اسلام. تهران: ۱۳۳۷، صفحات ۳۱ـ۲۰.

دهخدا، علی اکبر. شرح حال نابغهٔ شهیر ایران ابوریحان محمد بن احمد خوارزمی بیرونی. تهران: کتابخانهٔ طَهوری، ادارهٔ کل نگارش وزارت فرهنگ و هنر. چاپ دوم، ۱۳۵۲.

علامهٔ فقید علی اکبر دهخدا احتیاج به معرفی ندارد. کتاب دهخدا در مورد بیرونی یکی از بهترین کتابهایی است که به رشتهٔ تحریر در آمده است. این کتاب دارای لیست اسامی کتابهای بیرونی نیز میباشد. کلیهٔ سوالات و جوابهای مربوطه که بین بیرونی و ابن سینا رد و بدل شده در این کتاب به چاپ رسیده است.

سپهرُم، امیر مسعود. تاریخ برگُزیدگان وعده ای از مشاهیر ایران عرب. تهران: کتابفروشی زوار، ۱۳۴۱، صفحهٔ ۲۸۶.

شابی، علی. زندگینامهٔ بیرونی. تهران: وزارت فرهنگ و هنر، مرکز پژوهشهای مردمشناسی و فرهنگ عامه، ترجمهٔ پرویز اذکایی، بدون تاریخ چاپ.

این کتاب شامل سه بخش: سرگذشت بیرونی، روش تاریخنگاری بیرونی، زبان و ادب بیرونی میباشد و بوسیلهٔ علی الشابی به زبان عربی نوشته شده و بوسیلهٔ پرویز اذکایی به فارسی برگردانده شده است. این کتاب هم جالب و خواندنی میباشد.

صفا، ذبیح الله. تاریخ علوم عقلی در تمدن اسلامی تا اواسط قرن پنجم. تهران: ۱۳۳۶، صفحات ۲۸۱_۲۸۷.

صفا، ذبیح الله. احوال و آثار ابوریحان بیرونی. تهران: انتشارات ادارهٔ کل نگارش وزارت فرهنگ و هنر، چاپ دانشگاهٔ تهران، ۱۳۵۲.

دکتر ذبیح الله صفا، استاد ممتاز دانشگاه تهران، این کتاب را در سه بخش: گزارش حال، کتب و آثار و آراء و افکار بیرونی به رشتهٔ تحریر در آورده است. این کتاب جمعاً دارای ۲۵ فصل کوتاه میباشد. میتوان گفت که این کتاب یکی از بهترین منابع موجود و موثق در مورد بیرونی به شمار میآید.

قربانی، ابولقاسم. ریاضیدانان بزرگ ایران. تهران: ۱۳۵۰.

قربانی، ابولقاسم. بیرونی نامه: تحقیق در آثار استاد ابوریحان بیرونی ریاضیدان و منجم بزرگ ایران در سدهٔ چهارم و پنجم هجری. تهران: انتشارات انجمن آثار ملی شمارهٔ ۱۰۷، ۱۳۵۲.

فقط بخش اول این کتاب در دسترس ما قرار داشت. این بخش مشتمل بر: خلاصهٔ زندگینامهٔ بیرونی، شخصیت بیرونی از نظر متفکران شرق و غرب، بیرونی و احکام نجوم و منتخباتی از نوشته های بیرونی میباشد. بنا به گفتهٔ مولف موضوع اصلی کتاب حاضر تحقیق در آثار ریاضی استاد ابوریحان بیرونی بوده است.

کریمی، غلامعلی. آوای ابوریحان. اصفهان: چاپ دانشگاهٔ اصفهان، مهر ماهٔ ۱۳۵۲.

این کتاب بمناسبت هزارمین سال تولد استاد ابوریحان بیرونی بوسیلهٔ دکتر غلامعلی کریمی به دو زبان عربی و فارسی نوشته و در دانشگاهٔ اصفهان چاپ گردیده است. دکتر کریمی از بیرونی بعنوان سیاستمدار، ادیب، عالم، فیلسوف و یک انسان واقعی یاد مینماید.

معتضد، خسرو. ابوریحان بیرونی: داستان زندگی نابغهٔ ریاضی و نجوم و دیگر دانش ها. تهران: شرکت توسعهٔ کتابخانه های ایران، ۱۳۷۰.

در این کتاب سر گذشت بیرونی بیشتر در قالب داستان با نگرش به ریشه های تاریخی نوشته شده است. این کتاب در ۵۶ فصل کوتاه به قلم خسرو معتضد چاپ شده است. این نویسندهٔ معاصر کتابهایی در مورد زندگی چند دانشمند معروف و شخصیت های فرهنگی و سیاسی ایران به رشتهٔ تحریر در آورده است. این کتاب جنبهٔ علمی و تحقیقی نداشته ولی با قلمی گرم و جذاب داستان زندگی بیرونی را بیان نموده است. این کتاب به برای طبقات عام نوشته شده است.

نجفی، محمدعلی و خلیلی، مهیار. اندیشمند و انسان: ابوریحان بیرونی. تهران: چاپ نشریهٔ اندیشه، شهریور ماهٔ ۱۳۲۵.

همائی، جلال الدین. ابوریحان بیرونی: اختراعات و اکتشافات و فن نجوم. تهران: کتابخانهٔ طهوری، ۱۳۵۲.

استاد جلال الدین همایی این کتاب را در دو قسمت تحت عنوانهای ۱) احتراعات و اکتشافات ۲) ابوریحان و فن نجوم به رشته تحریر در آورده است. استاد همایی بعد از شمردن شانزده

مورد از اختراعات و اکتشافات بیرونی او را بعنوان کاشف آمریکا معرفی مینماید.

یادنامهٔ بیرونی، الف: مجموعه سخنرانیهای فارسی. تهران: چاپخانهٔ زر، ۱۳۵۳.

این کتاب مفصل در دو جلد چاپ شده است و دارای بیست و هشت مقالهٔ مفید در مورد زندگی بیرونی و تحقیقات علمی وی به چاپ رسیده است. این کتاب هم بمناسبت هزارمین سال تولد استاد بزرگ ایرانی تنظیم گردیده است. اکثر مولفین مقالات این کتاب خود از برجسته ترین دانشمندان ایرانی و خارجی متخصص در مورد بیرونی میباشند. این منبع یکی از بهترین کتابهای موجود در بارهٔ بیرونی به حساب می آید.

Conclusion

The main goal of this research was to bridge a gap in the history of scientific achievements. Reference sources on Biruni's scientific research and writings are almost unknown. Although we were able to find three existing bibliographies on Biruni and his works, unfortunately, none is complete or up to date. Two of them were published in 1973 and another one was written in 1982. This means that the most recent bibliography on Biruni is nearly a quarter of a century old. Thus, the purpose of this research was to bring together in a single volume all published materials in English and Persian that are related to Biruni and his research agenda.

In the first part of this Bio-Bibliography we provided the reader with a well-balanced introduction to Biruni. In this part we dealt with his time, his world, and his life. In Part II, we were able to provide full titles of all of the 183 books written by Biruni. The present project is the first to present these titles in Arabic, Persian, transliteration, and English. Comparatively speaking, our annotations of these books, although sometimes very brief, must be considered as the most comprehensive available. In addition to traditional library research, again, for the first time, we inspected the Internet and identified the most important Web sites on Biruni and his works. Parts III-V contain all of the references that we could find on him and his works.

We are happy to report that we have found a full-length film and a doctoral dissertation on Biruni and his works. We believe that our bibliography is not only up to date, but it is also the most complete available. We are hoping that we have made a contribution toward understanding of one of the most important scholars with authoritative works of outstanding quality in the history of mankind.

Appendix A

Names of Biruni's 183 books in Arabic with Accents

١ـ المسائل المُفيده وَ الجَوباتُ السَديده.

٢ـ إبطالُ البُهتان بايراد البُرهان اعمالِ الخوارزمى فى زَيجه.

٣ ـ كِتاب الوَساطه بينهُما (الخَوارزمى و الاهوازى).

٤ـ تكميلُ زَيج حَبَش بالعِلل وَ تَهذيبُ أعماله مِن الزَلَل.

٥ـ جَوامعُ الموجُود لِخواطِر الهُنُود فى حسابِ التَنجيم.

٦ـ تَهذيبُ زَيج الاركند.

٧ـ كِتاب مَقاليدُ علم الهَيئت وَ ما يُحدثُ فى بسيط الكُره.

٨ـ خَيالُ الكُسوفَين عِند الهِند.

٩ـ فى أمرُ المُتحَن وَ تَبصيرُ ابن كيسوم المُفتَتن.

١٠ـ اختِلافُ الاقاويل لاستِخراج التَحاويل.

١١ـ مَقاله فى التَحليلُ وَ التَقطيع للتعديل.

١٢ـ فى تَهذيبُ الطرُق المُحتاج اليها فى اِستخراج هيئت الفلك عند المواليد و تحاويل السنين وَ غَيرها مِن الاوقات.

١٣ـ مِفتاحُ علم الهَيئت.

١٤ـ تَهذيبُ فُصول الفَرغانى.

١٥ـ كِتاب فى افرادُ المَقال فى امر الاظَلال.

١٦ـ كِتاب فى استعمالُ دوائر السُموت لاستخراج مَراكِز البُيُوت.

١٧ـ مَقاله فى طالعُ قُبه الارض وَ حالاتُ الثوابت ذوات العُرُوض.

١٨ــ مَقاله صَغيره في اعتبارُ مِقدار اللّيل وَ النّهار في جَميع الارض لتعريف كونِ السّنه يوماً تحت القطب بغير تَشكيل.

١٩ــ كِتاب تَحديد نَهايات الاماكن لتَصحيح مَسافات المساكن.

٢٠ــ كِتاب تَهذيب الاقوال في تَصحيح العُروض و الاطوال.

٢١ــ كِتاب تَصحيفُ المنقول مِن العَرض و الطُول.

٢٢ــ مَقاله في تَصحيحُ الطُول وَ العَرض لِمساكن المعمُور مِن الارض.

٢٣ــ مَقاله في تعيين البَلَد مِن العَرض وَ الطول.

٢٤ــ مَقاله في اِستخراج قَدر الارض بِرَصَد انحِطاط الأُفُق عن قُلل الجِبال.

٢٥ــ في غُرُوب الشَمس عِند مَنارات الاسكندريه.

٢٦ــ في اختلافُ الواقع في تَقسيم الاقاليم.

٢٧ــ في اختلافُ ذوي الفضل في اِستخراج العُروض و الميل.

٢٨ــ كِتاب الاجوبه وَ الاسئله لتَصحيح سَمتِ القِبلَه.

٢٩ــ إيضاحُ الادِله على كيفيت سمت القِبله.

٣٠ــ تَهذيب شُرُوط العَمَل لتَصحيح سُمُوت القِبلَه.

٣١ــ في تَقويم القِبله في بُست بتَصحيح طولها و عَرضها.

٣٢ــ في الانبِعاث لِتَصحيح القِبله.

٣٣ــ تَلافي عوارض الزّلّه في كِتاب دلائل القِبلَه.

٣٤ــ تَذكره في الحِساب وَ العَد بارقام السِند وَ الهِند.

٣٥ــ في اِستخراجُ الكِعاب وَ اضلاع ماورائه من مَراتِب الحِساب.

٣٦ــ كيفيتُ رُسُوم الهِند في تعلُم الحِساب.

٣٧ــ في انّ رأى العَرب في مَراتِب العَدَد اَصوبُ من رأى الهِند فيها.

٣٨ــ في رشيكاتُ الهِند.

٣٩ـ فی سَکلب الاعداد.

٤٠ـ تَرجُمه ما فی بَرَاهَم سِدهاند مِن طُرُق الحِساب.

٤١ـ مَنصُوباتُ الضَرب.

٤٢ـ تَجریدُ الشُعاعاتِ وَ الانوار عن الفَضائح المُدونه فی لاسفار.

٤٢ـ مَقاله فی تَحصیلُ الشُعاعات بابعاد الطُرُق عَن الساعات.

٤٤ـ فی مَطرحُ الشُعاع الثابتاً علی تَغیَرُ البقاع.

٤٥ـ تَمهیدُ المُستَقَرق لِتَحقیق معنی المَمر.

٤٦ـ کِتاب فی استیعاب الوُجُوه امُکنه فی صَنعت الاُسطُرلاب.

٤٧ـ فی تَسهیل التَصحیح الاُسطُرلابی وَ العمل بمُرکباته من الشُمالی وَ الجُنوبی.

٤٨ـ فی تَسطیحُ الصُور وَ تَبطیحُ الکور.

٤٩ـ فی ما اُخرج ما فی قوه الاُسطُرلاب.

٥٠ـ فی استعمال الاُسطُرلاب الکُروی.

٥١ـ مَقاله فی تَعبیرُ المیزان لِتَقدیر الزَمان.

٥٢ـ فی تَحصیلُ الآن من الزَمان عند الهِند.

٥٣ـ تَذکره فی الارشاد الی صَوم النَصارا وَ الاعیاد.

٥٤ـ فی الاعتِذار عما سَبَق لی فی تاریخ الاسکندر

٥٥ـ فی تَکمیلُ حکایات عَبدالملک الطبیبِ البُستی فی مبدأ العالم وَ انتهائه.

٥٦ـ مَقاله فی دَلالتُ الآثارِ العُلویه عَلَی الاحداثِ السُفلیه.

٥٧ـ فی اِبطال ظُنون فاسِده خَطرَته عَلَی قُلوب بَعض الاطباء فی أمر الکواکب الحادِثه فی الجوَ.

٥٨ـ مَقاله فی الکلام عَلَی کواکب ذَوات الأذناب وَ النوائب.

٥٩ـ مَقاله فی مضیاتُ الجو الحادِثه فی العَلوُ.

٦٠ـ مَقاله فی تَصَفحُ کلامِ ابی سَهل القُوهی فِی الکواکِبِ المُنقضه.

٦١_ كِتاب فی تَحقیقُ مَنازِل القَمر.

٦٢_ فی الفَحص عن نَوادر اَبی حَفص عُمَر ابن الفَرُخان.

٦٣_ مَقاله فی النسبُ التی بین الفِلِزات وَ الجَواهر فی الحَجم.

٦٤_ مَقاله فی اِستخراجُ الاوتار فی الدائره بخواص الخَط المُنحنی فیها.

٦٥_ تَذکره فی المساحت للمُسافر المقوی.

٦٦_ مَقاله فی نَقلُ خواص الشِکل القطاع الا ما یغُنی عَنه.

٦٧_ مَقاله فی أنَ لوازم تَجزی المقادیر الاَنهایه قریبت من أمر الخَطین الذین یَقرَبان وَ لا یَلتَقیان فی الاستِبعاد.

٦٨_ مَقاله فی صِفتُ أسباب السُخُونه الموجوده فی العام وَ اختلاف اُصول السنه.

٦٩_ مَقاله فی البَحثُ عن الطَریقت المُتعارفه المذکوُره فی کِتاب الآثار العُلویه.

٧٠_ المسائل البَلخیه فی المعنی المُتعلقه بانکِسار الصِناعت.

٧١_ الجَوابات عَن المسائل الوارده مِن مُنجمی الهِند.

٧٢_ الجوابات عَن المسائل العَشر الکِشمیریه.

٧٣_ کتاب التَفهیم لاوائل صِناعت التَنجیم.

٧٤_ مَقاله فی تَقسیط القُوه و الدلالات بَین اجزاء البُیوت الاثنا عَشر.

٧٥_ مَقاله فی حِکایت طَریق الهِند فی اِستخراج العُمر.

٧٦_ مَقاله فی سَیر سَهمی السَعاده وَ الغَیب.

٧٧_ فی الارشاد علی تَصحیح المبادی اِشتَمَل علی النَمُوذرات.

٧٨_ مَقاله فی تبین رأی بَطلمیوس فی سالخُداه.

٧٩_ تَرجُمه کِتاب المواليد الصَغیره لبراهیمیهرا.

٨٠_ تَرجُمه قِصه وامِق وَ عَذرا.

٨١_ تَرجُمه حَدیث قاسم السُرُور وَ عین الحَیات.

۸۲ـ تَرجُمه حَديث أورمزديا وَ مِهريار.

۸۳ـ تَرجُمه حَديث صَنَمَى الباميان.

۸٤ـ تَرجُمه حَديث دائمِه وَ كِرامى دُخت، بِنت جَهل الوادى.

۸۵ـ ترجمه حَديث نيلوفَر فى قِصه دَبيستى وَ بَربهاكر.

۸٦ـ تِرجُمه قافيُه الالف مِن الاتمام فى شِعر اَبى تمام.

۸۷ـ تَرجُمه مَقاله فى الاستِبحار فى قَد الاشجار.

۸۸ـ تَرجُمه تَحصيل الراحَه بتَصحيح المساحَه.

۸۹ـ تَرجُمه التَحذير من قِبل التُرک.

۹۰ـ تَرجُمه القُرعه المُصرَحَه بالعواقِب.

۹۱ـ تَرجُمه القُرعه المُثَمَنه لاستِنباط الضَمائر المُخَمَنه وَ شَرح مَزامير القُرعه المُثَمَنه.

۹۲ـ تَرجُمه كَلَب ياره وَ هو مَقاله للهِند فى الامراض الَتى تَجرى مَجرا العُفُونه.

۹۳ـ كِتاب فى تَحقيق ما للهِند من مَقاله مَقبوله فى العَقل اومرذوله.

۹٤ـ مَقاله فى عِلت علامات البُرُوج فى الزيجات من حروف الجُمَل.

۹۵ـ كلام فى المُستَقر وَ المُستودَع.

۹٦ـ مَقاله فى بَسديو الهِند عِند مَجيئه الادنا.

۹۷ـ تَرجُمه كِتاب شامِل فى الموجُوداتِ المَحسوسَه وَ المَعقوُله.

۹۸ـ تَرجُمه كِتاب بَاتَنجَل فى الخَلاص مِن الارتِباک.

۹۹ـ التَنبيه عَلى صِناعات تَمويه وَ هى احكام النُجوم.

۱۰۰ـ تَنويرُ المِنهاج الى تَحليل الازياج.

۱۰۱ـ التَطبيق الى تَحقيقِ حَركت الشَمس.

۱۰۲ـ البُرهان المُنير فى اَعمال التَسيير.

۱۰۳ـ كِتاب تَنقيحُ التَواريخ وَ اَمثالُ ذلک.

١٠٤ــ القانُون المسعوُدى.

١٠٥ــ الآثار الباقيه عَن القُرون الخاليه.

١٠٦ــ الارشاد اِلى ما يُدرکُ ولا يُنالُ مِن الأبعاد.

١٠٧ــ الکِتابُ فی المکاييل وَ الموازين وَ شرايط الطَيار وَ الشواهين.

١٠٨ــ جَمع الطُرُق السائره فی مَعرفت أوتارِ الدائره.

١٠٩ــ تَصور أمر الفَجر وَ الشَفَق فی جَهتی الشَرق وَ الغَرب من الاُفُق.

١١٠ــ تَکميل صِناعه التَسطيح.

١١١ــ جَلاءُ الأذهان فی زَيج البتانی.

١١٢ــ تَحديدُ المعمُوره وَ تَصحيحُها فی الصوَره.

١١٣ــ عِلَل زَيج جَعفَر المُکنی بأبی مَعشر.

١١٤ــ کِتاب فی السُمُوت.

١١٥ــ کِتاب فی علتُ تَنصيف التَعديل عِند اصحابِ السِند هِند.

١١٦ــ کِتاب فی تَصحيحُ کِتابِ ابراهيم بن سِنان فی تَصحيح اختِلافِ الکواکب العُلويه.

١١٧ــ رساله فی براهينُ أعمال حَبَش بجَدول التَقويم.

١١٨ــ کِتاب فی تَصحيح ما وقع لأبی جَعفَر الخازِن من السَهو فی زَيج الصَفائح.

١١٩ــ رساله فی مَجازاتُ دوائر السُمُوت فی الأسطُرلاب.

١٢٠ــ رساله فی جَدول الدقائق.

١٢١ــ رساله فی البُراهين عَلی عَمَل مُحَمد بن الصباح فی امتِحانِ الشَمس.

١٢٢ــ رساله فی الدوائر الَتی تَحُدُ الساعات الزَمانيه.

١٢٣ــ رساله فی البُرهان علی عَمَل حَبَش فی مَطالع السمت فی زَيجه.

١٢٤ــ رساله فی مَعرفَت القُسی الفَلَکيه بطَريق غَير طَريق النِسبَت المولفه.

١٢٥ــ رساله فی حَل شُبهَت عَرَضَت فی الثاثه عَشر من کتاب الاُصُول.

١٢٦_ كِتاب في مَبادى الهندسه.

١٢٧_ كتاب في رُسُوُم الحَركات في الاشياء ذوات الوَضع.

١٢٨_ كِتاب في سُكُوُن الارضِ اَوحَرَكاتُها.

١٢٩_ كِتاب في التَوَسُط بَينَ ارسطوُطاليس وَ جالينوس في المُحركِ الاول.

١٣٠_ رساله في دَلالت اللفظ على المعنى.

١٣١_ رساله في سَبَبُ بَرد ايام العَجُوز.

١٣٢_ رساله في عِلتُ التَربيه التي تُستَعملُ في احكام النُجُوم

١٣٣_ رساله في آداب صُحبت المُلوك.

١٣٤_ رساله في قوانين الصَناعت.

١٣٥_ رساله في دَستوُرُ الخَط.

١٣٦_ رساله في الغَزَليات الشَمسيه.

١٣٧_ الرسالهُ النَرجسيه.

١٣٨_ الرسالهُ المعنونه بِمَن وَ عَن.

١٣٩_ استيعاب في تَسطيح الكُره.

١٤٠_ الزيجُ العلائي.

١٤١_ الزيجُ المسعودى.

١٤٢_ كِتاب تَسطيحُ الكُره.

١٤٣_ كِتاب الدُرر في سَطح الأكر.

١٤٤_ اختِصار المِجسطي.

١٤٥_ كِتاب في لَوازِمُ الحَركتين.

١٤٦_ كِتاب الاستِشهاد باختِلاف الارصاد.

١٤٧_ مسائل سأل عَنها ابن سينا.

١٤٨ـ كتاب الشُمُوسُ الشافيه للنفُوس.

١٤٩ـ ارشاد في احكام النُجُوم.

١٥٠ـ كِتاب تاريخُ ايام السُلطان مَحمُود وَ أخبار أبيه.

١٥١ـ في صَفيحةُ التسيير.

١٥٢ـ كِتاب المدخَل الى عِلم النُجُوم وَ عَلى حُدوُد النُحوس وَ المواضع القاطه بالتَجارب.

١٥٣ـ كِتاب العجايبُ الطَبيعت وَ الغرائب الصناعت.

١٥٤ـ كِتاب تَقاسيمُ الاقاليم.

١٥٥ـ كِتاب في اعتبارُ مِقدار اللَيل وَ النَهار بطَريقين تَبعُدُ عَن مُوضعات المُنجمين وَ ألقابهم.

١٥٦ـ الجَماهِر في الجَواهِر.

١٥٧ـ كِتاب نُزهَت النُفوس وَ الافكار في خَواص المواليد الثلاثه المعادِن وَ النَبات وَ الحَجار.

١٥٨ـ كِتاب الصَيدلة في الطِب.

١٥٩ـ كِتاب الحِجاَر.

١٦٠ـ كِتاب المُسامرهُ في اَخبار خوارَزم.

١٦١ـ ترجمة اخبارُ المُقَنع.

١٦٢ـ كِتاب في اخَبار المبَيَضه وَ القَرَامطِه.

١٦٣ـ كتاب شَرحُ شِعر أبي تَمام.

١٦٤ـ كِتاب التَعَلُل بحالات الوَهم في مَعاني نَظم عُلى الفَضل.

١٦٥ـ كِتاب مُختارُ الاشعار وَ الآثار.

١٦٦ـ مَجمُوعه مِن الاشعار.

١٦٧ـ كتاب المُعَنون بالدَستُور.

١٦٨ـ رساله في فِهرست كُتُب مُحَمَد بن زَكريا الرازي.

١٦٩ـ حِكايت الاله المُسَمات السُدس الفَخري.

۱۷۰_ رياضَت الفِكر وَ العَقل.

۱۷۱_ اجوبَت المسائل الخوارزميه.

۱۷۲_ في الاعتراض عَلى كِتاب ابن سينا في حُجت الحَق.

۱۷۳_ مَجمُوعه مِن الحِكم.

۱۷٤_ ترجُمه كِتابُ سامخيه.

۱۷٥_ تَرجُمه كِتاب في أُصُوُل الهِندسه لأُقليدُس الى لُغت الهِند.

۱۷٦_ تَرجُمه كِتاب المَجسطي لبَطلَميوُس الى لُغت الهِند.

۱۷۷_ تَرجُمه كِتاب في الأسطُرلاب لابي رَيحان الى لُغت الهِند.

۱۷۸_ رساله أبي نصر في جواب مسائل الهندسه.

۱۷۹_ فَصل مِن كِتاب لأبي نَصر في كُريت السَماء.

۱۸۰_ كِتابُ ابي الرَيحان الى ابي سَعيد.

۱۸۱_ غُره الزيجات أو غُره الازياج وَ مَعنى زَيج بجاى نند.

۱۸۲_ كِتاب اللَمَعات.

۱۸۳_ تَرجُمه ما في پوليسا سَدهانتا.